MW01595807

Cram101 Textbook Outlines to accompany:

Essentials of Biology

Sylvia S. Mader, 2nd Edition

With Cram101.com online, you also have access to extensive reference material.

You will nail those essays and papers. Here is an example from a Cram101 Biology text:

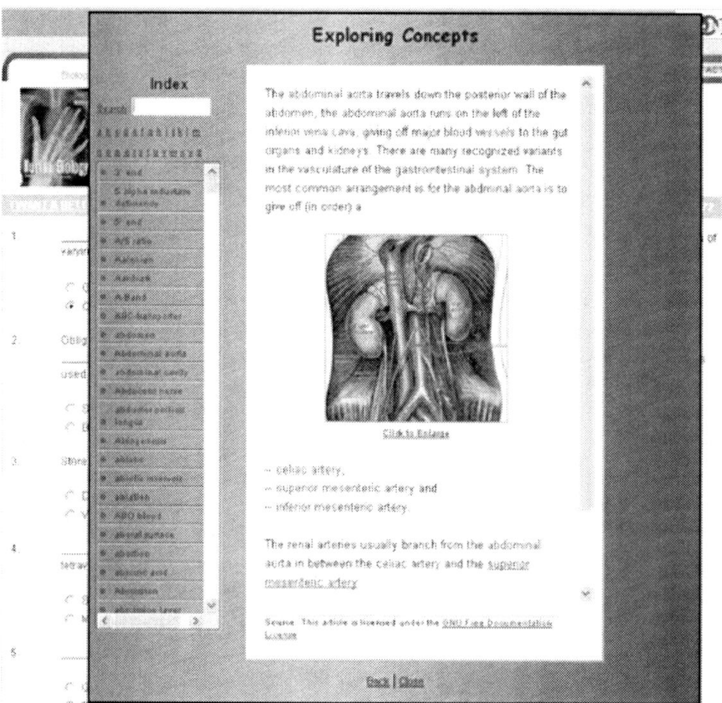

Visit **www.Cram101.com**, click Sign Up at the top of the screen, and enter DK73DW8077 in the promo code box on the registration screen. Access to www.Cram101.com is normally $9.95 per month, but because you have purchased this book, your access fee is only $4.95 per month, cancel at any time. Sign up and stop highlighting textbooks forever.

Learning System

Cram101 Textbook Outlines is a learning system. The notes in this book are the highlights of your textbook, you will never have to highlight a book again.

How to use this book. Take this book to class, it is your notebook for the lecture. The notes and highlights on the left hand side of the pages follow the outline and order of the textbook. All you have to do is follow along while your instructor presents the lecture. Circle the items emphasized in class and add other important information on the right side. With Cram101 Textbook Outlines you'll spend less time writing and more time listening. Learning becomes more efficient.

Cram101.com Online

Increase your studying efficiency by using Cram101.com's practice tests and online reference material. It is the perfect complement to Cram101 Textbook Outlines. Use self-teaching matching tests or simulate in-class testing with comprehensive multiple choice tests, or simply use Cram's true and false tests for quick review. Cram101.com even allows you to enter your in-class notes for an integrated studying format combining the textbook notes with your class notes.

 ISBN(s): 9781428891531.

Essentials of Biology
Sylvia S. Mader, 2nd

CONTENTS

Leukemia inhibitory factor	Leukemia inhibitory factor an interleukin 6 class cytokine, is a chemical in cells that affects their growth and development. Leukemia inhibitory factor derives its name from its ability to induce the terminal differentiation of myeloid leukaemic cells. Other properties attributed to the cytokine include: the growth promotion and cell differentiation of different types of target cells, influence on bone metabolism, cachexia, neural development, embryogenesis and inflammation.
Cell	The Cell is the structural and functional unit of all known living organisms. It is the smallest unit of an organism that is classified as living, and is often called the building block of life. Some organisms, such as most bacteria, are unicellular (consist of a single Cell.)
Multicellular organisms	Multicellular organisms are organisms consisting of more than one cell, and having differentiated cells that perform specialized functions in the organism. Most life that can be seen with the naked eye is multicellular, as are all members of the kingdoms Plantae and Animalia (except for specialized organisms such as Myxozoa in the case of the latter.) Early life was most probably single celled and multicellularity has appeared dozens of times in the history of Earth.
Population	In biology, a population is the collection of inter-breeding organisms of a particular species; in sociology, a collection of human beings. Individuals within a population share a factor may be reduced by statistical means, but such a generalization may be too vague to imply anything. Demography is used extensively in marketing, which relates to economic units, such as retailers, to potential customers.
Chromosome	A Chromosome is an organized structure of DNA and protein that is found in cells. It is a single piece of coiled DNA containing many genes, regulatory elements and other nucleotide sequences. Chromosome s also contain DNA-bound proteins, which serve to package the DNA and control its functions.
Decomposers	Decomposers are organisms that consume dead or decaying organisms, and, in doing so, carry out the natural process of decomposition. Like herbivores and predators, Decomposers are heterotrophic, meaning that they use organic substrates to get their energy, carbon and nutrients for growth and development. Decomposers use deceased organisms and non-living organic compounds as their food source.
Horse murders	The Horse murders scandal was a form of insurance fraud in the United States in which expensive horses, many of them show jumpers, were insured against death, accident and then killed to collect the insurance money. It is not known how many horses were murdered between the mid 1970s and the mid-1990s, when a Federal Bureau of Investigation (FBI) investigation brought the horse killings to light, but the number is thought to be well over 50, and may have been as high as 100. In addition, in 1977, the heiress Helen Brach disappeared and was presumed by law enforcement agents to have been murdered by the perpetrators of these crimes, because she threatened to report their criminal activity to authorities; continuing investigations into Brach's death began to uncover the insurance fraud in the 1990s.
Metabolism	Metabolism is the set of chemical reactions that occur in living organisms in order to maintain life. These processes allow organisms to grow and reproduce, maintain their structures, and respond to their environments. Metabolism is usually divided into two categories.

Photosynthesis	Photosynthesis is a process that converts carbon dioxide into organic compounds, especially sugars, using the energy from sunlight. Photosynthesis occurs in plants, algae, and many species of Bacteria, but not in Archaea. Photosynthetic organisms are called photoautotrophs, since it allows them to create their own food.
Acid	An acid is traditionally considered any chemical compound that, when dissolved in water, gives a solution with a hydrogen ion activity greater than in pure water, i.e. a pH less than 7.0. That approximates the modern definition of Johannes Nicolaus Brønsted and Martin Lowry, who independently defined an acid as a compound which donates a hydrogen ion (H^+) to another compound (called a base.) Common examples include acetic acid and sulfuric acid (used in car batteries.)
Dentition	Dentition is the development of teeth and their arrangement in the mouth. All mammals except the monotremes, the xenarthrans, the pangolins, and the cetaceans have up to four distinct types of teeth, with a maximum number for each. These are the incisor (cutting), the canine, the premolar, and the molar (grinding.)
Gene	A Gene is the basic unit of heredity in a living organism. All living things depend on Gene s. Gene s hold the information to build and maintain their cells and pass Gene tic traits to offspring.
Edward	The Edward mango is a monoembryonic mango cultivar grown predominantly in Florida. It is considered by many to be among the finest tasting mangoes in the world; however, its poor yields have restrained the Edward from developing into a commercially significant variety. The Edward was first propagated in the 1920s by Edward Simmonds of the Plant Introduction Garden in Miami, Florida and is believed to be a hybrid cross of Haden and Carabao mango cultivars.
Escherichia	Escherichia is a genus of Gram-negative, non-spore forming, facultatively anaerobic, rod-shaped bacteria from the family Enterobacteriaceae. Inhabitants of the gastrointestinal tracts of warm-blooded animals, Escherichia species provide a portion of the microbially-derived vitamin K for their host. While many Escherichia are harmless commensals, particular strains of some species are human pathogens, and are known as the most common cause of urinary tract infections, significant sources of gastrointestinal disease, ranging from simple diarrhea to dysentery-like conditions, as well as a wide-range of other pathogenic states.
Escherichia coli	Escherichia coli , is a Gram negative bacterium that is commonly found in the lower intestine of warm-blooded organisms . Most E. coli strains are harmless, but some, such as serotype O157:H7, can cause serious food poisoning in humans, and are occasionally responsible for costly product recalls. The harmless strains are part of the normal flora of the gut, and can benefit their hosts by producing vitamin K_2, or by preventing the establishment of pathogenic bacteria within the intestine.
Archaea	The Archaea are a group of single-celled microorganisms. A single individual or species from this domain is called an archaeon (sometimes spelled 'archeon'.) They have no cell nucleus or any other organelles within their cells.

Bacteria

The Bacteria are a large group of unicellular microorganisms. Typically a few micrometres in length, Bacteria have a wide range of shapes, ranging from spheres to rods and spirals. Bacteria are ubiquitous in every habitat on Earth, growing in soil, acidic hot springs, radioactive waste, water, and deep in the Earth's crust, as well as in organic matter and the live bodies of plants and animals.

DNA

Deoxyribonucleic acid (DNA) is a nucleic acid that contains the genetic instructions used in the development and functioning of all known living organisms and some viruses. The main role of DNA molecules is the long-term storage of information. DNA is often compared to a set of blueprints or a recipe, or a code, since it contains the instructions needed to construct other components of cells, such as proteins and RNA molecules.

Class

In biological classification, Class is

· a taxonomic rank. Other well-known ranks are life, domain, kingdom, phylum, order, family, genus, and species, with Class fitting between phylum and order. As for the other well-known ranks, there is the option of an immediately lower rank, indicated by the prefix sub-: subclass .

· a taxonomic unit, a taxon, in that rank. In that case the plural is classes

The composition of each Class is determined by a taxonomist. Often there is no exact agreement, with different taxonomists taking different positions. There are no hard rules that a taxonomist needs to follow in describing a Class, but for well-known animals there is likely to be consensus. For example, dogs are usually assigned to the Class Mammalia; in the phylum Chordata (animals with notochords); in the order Carnivora (mammals that eat meat.)

Family

In biological classification, Family is

· a taxonomic rank. Other well-known ranks are life, domain, kingdom, phylum, class, order, genus, and species, with Family fitting between order and genus. As for the other well-known ranks, there is the option of an immediately lower rank, indicated by the prefix sub-: subfamily .

· a taxonomic unit, a taxon, in that rank. In that case the plural is families

Example: 'Walnuts and Hickories belong to the Walnut Family.

What does and does not belong to each Family is determined by a taxonomist. Similarly for the question if a particular Family should be recognized at all. Often there is no exact agreement, with different taxonomists each taking a different position.

Fungi

A fungus is a eukaryotic organism that is a member of the kingdom fungi . The fungi are a monophyletic group, also called the Eumycota , that is phylogenetically distinct from the structurally similar slime molds (myxomycetes) and water molds (oomycetes.) The fungi are heterotrophic organisms possessing a chitinous cell wall, with most species growing as multicellular filaments called hyphae forming a mycelium; some species also grow as single cells.

Genus

A Genus is a low-level taxonomic rank used in the classification of living and fossil organisms, and also any taxonomic unit (taxon) of that rank. The binomial name of every species is formed from a Genus name (with a capital initial), followed by the species name, both normally written in italics.

The term comes from Latin Genus 'descent, family, type, gender' , cognate with Greek: γῖνος - genos, 'race, stock, kin'.

Kingdom

In biological taxonomy, Kingdom or regnum is a taxonomic rank in either (historically) the highest rank, or (in the new three-domain system) the rank below domain. Each Kingdom is divided into smaller groups called phyla (or in some contexts these are called 'divisions'.) Currently, many textbooks from the United States use a system of six kingdoms (Animalia, Plantae, Fungi, Protista, Archaea, Bacteria) while British and Australian textbooks may describe five kingdoms (Animalia, Plantae, Fungi, Protista, and Prokaryota or Monera.)

Order

In scientific classification used in biology, the Order is

- a taxonomic rank used in the classification of organisms. Other well-known ranks are life, domain, kingdom, phylum, class, family, genus, and species, with Order fitting in between class and family. An immediately higher rank, superorder, may be added directly above Order, while suborder would be a lower rank.
- a taxonomic unit, a taxon, in that rank. In that case the plural is orders .

The Latin suffix -formes meaning 'having the form of' is used for the scientific name of orders of birds and reptiles, but not for those of mammals and invertebrates.

The Order as a distinct rank of biological classification having its own distinctive name (and not just called a higher genus (genus summum)) was first introduced by a German botanist Augustus Quirinus Rivinus in his classification of plants . Carolus Linnaeus was the first to apply it consistently to the division of all three kingdoms of Nature (minerals, plants, and animals) in his Systema Naturae (1735, 1st. Ed.).

Species

In biology, a Species is:

- a taxonomic rank (the basic rank of biological classification) or
- a unit at that rank

There are many definitions of what kind of unit a Species is (or should be.) A common definition is that of a group of organisms capable of interbreeding and producing fertile offspring, and separated from other such groups with which interbreeding does not (normally) happen. Other definitions may focus on similarity of DNA or morphology. Some Species are further subdivided into sub Species , and here also there is no close agreement on the criteria to be used.

Technology

Technology is a broad concept that deals with an animal species' ethology or behavior of usage and of knowledge of tools and crafts, and how it affects the animal species' ability to control and adapt to its environment. Technology is a term with origins in the Greek 'technologia', 'τεχνολογΐ¯α' -- 'techne', 'τΐχνη' and 'logia', 'λογΐ¯α' ('saying'.) However, a strict definition is elusive; 'Technology' can refer to material objects of use to humanity, such as machines, hardware or utensils, but can also encompass broader themes, including systems, methods of organization, and techniques.

Charles Robert Darwin

Charles Robert Darwin FRS (12 February 1809 - 19 April 1882) was an English naturalist who realised and presented compelling evidence that all species of life have evolved over time from common ancestors, through the process he called natural selection. The fact that evolution occurs became accepted by the scientific community and much of the general public in his lifetime, while his theory of natural selection came to be widely seen as the primary explanation of the process of evolution in the 1930s, and now forms the basis of modern evolutionary theory. In modified form, Darwin's scientific discovery is the unifying theory of the life sciences, providing logical explanation for the diversity of life.

NADPH

Nicotinamide adenine dinucleotide phosphate ($NADP^+$, in older notation triphosphopyridine nucleotide, TPN) is used in anabolic reactions, such as lipid and nucleic acid synthesis, which require NADPH as a reducing agent.

NADPH is the reduced form of $NADP^+$, and $NADP^+$ is the oxidized form of NADPH. NADP+ differs from NAD+ by the presence in NADP+ of an additional phosphate group on the 2' position of the ribose ring that carries the adenine moiety.

In chloroplasts, NADP is reduced by ferredoxin-NADP+ reductase in the last step of the electron chain of the light reactions of photosynthesis.

Animals

Animals are a major group of mostly multicellular, eukaryotic organisms of the kingdom Animalia or Metazoa. Their body plan eventually becomes fixed as they develop, although some undergo a process of metamorphosis later on in their life. Most Animals are motile, meaning they can move spontaneously and independently.

Plants

Plants are living organisms belonging to the kingdom Plantae. They include familiar organisms such as trees, herbs, bushes, grasses, vines, ferns, mosses, and green algae. About 350,000 species of Plants, defined as seed Plants, bryophytes, ferns and fern allies, are estimated to exist currently.

Group

In naming cultivated plants, a Group is a formal classification category, under the International Code of Nomenclature for Cultivated Plants (ICNCP):

> ICNCP Art. 3.1: 'a formal category for assembing cultivars, individual plants or assemblages of plants on the basis of defined similarity' .

The term 'Group' (with a capital G) was introduced in the 2004 ICNCP, replacing the 'Cultivar-Group' of the 1995 ICNCP.

A Group is united by some common trait; for example there may be a Group of yellow-flowering cultivars, a Group of cultivars with variegated leaves, a Group of cultivars resistant to a particular disease, etc.

Leaf

In botany, a Leaf is an above-ground plant organ specialized for photosynthesis. For this purpose, a Leaf is typically flat (laminar) and thin. There is continued debate about whether the flatness of leaves [[Natural selection | evolved] to expose the chloroplasts to more light or to increase the absorption of carbon dioxide.

Biochemistry

Biochemistry is the study of the chemical processes in living organisms. It deals with the structure and function of cellular components such as proteins, carbohydrates, lipids, nucleic acids and other biomolecules.

Although there are a vast number of different biomolecules many are complex and large molecules (called polymers) that are composed of similar repeating subunits (called monomers.)

Biodiversity

Biodiversity is the variation of life forms within a given ecosystem, biome, or for the entire Earth. biodiversity is often used as a measure of the health of biological systems. The biodiversity found on Earth today consists of many millions of distinct biological species, which is the product of nearly 3.5 billion years of evolution.

Chemical	A chemical substance is a material with a specific chemical composition. A common example of a chemical substance is pure water; it has the same properties and the same ratio of hydrogen to oxygen whether it is isolated from a river or made in a laboratory. Some typical chemical substances are diamond, gold, salt (sodium chloride) and sugar (sucrose.)
Leukemia inhibitory factor	Leukemia inhibitory factor an interleukin 6 class cytokine, is a chemical in cells that affects their growth and development. Leukemia inhibitory factor derives its name from its ability to induce the terminal differentiation of myeloid leukaemic cells. Other properties attributed to the cytokine include: the growth promotion and cell differentiation of different types of target cells, influence on bone metabolism, cachexia, neural development, embryogenesis and inflammation.
Ascaris	Ascaris is a genus of parasitic nematode worms known as the giant intestinal roundworms. One species, A. suum, typically infects pigs, while another, A. lumbricoides, affects human populations, typically in sub-tropical and tropical areas with poor sanitation. A. lumbricoides is the largest intestinal roundworm and is the most common helminth infection of humans worldwide, an infection known as ascariasis.
Ascaris lumbricoides	Ascaris lumbricoides is the member of the Ascaris family responsible for the disease ascariasis. It can reach a length of up to 35 cm. Ascaris lumbricoides, or 'roundworm', infections in humans occur when an ingested infective egg releases a larval worm that penetrates the wall of the duodenum and enters the bloodstream.
Edward	The Edward mango is a monoembryonic mango cultivar grown predominantly in Florida. It is considered by many to be among the finest tasting mangoes in the world; however, its poor yields have restrained the Edward from developing into a commercially significant variety. The Edward was first propagated in the 1920s by Edward Simmonds of the Plant Introduction Garden in Miami, Florida and is believed to be a hybrid cross of Haden and Carabao mango cultivars.
Nereis	Nereis is a genus of polychaete worms in the family Nereidae. It comprises many species, most of which are marine, including the sandworm (Nereis virens) and the common clam worm (Nereis succinea.) Nereis possess setae and parapodia for locomotion.
Punnett square	The Punnett square is a diagram that is used to predict the outcome of a particular cross or breeding experiment. It is named after Reginald C. Punnett, who devised the approach, and is used by biologists to determine the probability of an offspring having a particular genotype. The Punnett square is a summary of every possible combination of one maternal allele with one paternal allele for each gene being studied in the cross.
Trisomy	A Trisomy is a genetic abnormality in which there are three copies, instead of the normal two, of a particular chromosome. Most organisms that reproduce sexually have pairs of chromosomes in each cell, with one chromosome inherited from each parent. In such organisms, a process called meiosis creates cells called gametes (eggs or sperm) that have only one set of chromosomes.

Salmonella	Salmonella is a genus of rod-shaped, Gram-negative, non-spore forming, predominantly motile enterobacteria with diameters around 0.7 to 1.5 Âµm, lengths from 2 to 5 Âµm, and flagella which project in all directions (i.e. peritrichous.) They are chemoorganotrophs, obtaining their energy from oxidation and reduction reactions using organic sources and are facultative anaerobes; most species produce hydrogen sulfide, which can readily be detected by growing them on media containing ferrous sulfate, such as TSI. Most isolates exist in two phases; phase I is the motile phase and phase II the non-motile phase. Cultures that are non-motile upon primary culture may be swithched to the motile phase using a Craigie tube.
Ion	An ion is an atom or molecule where the total number of electrons is not equal to the total number of protons, giving it a net positive or negative electrical charge. Since protons are positively charged and electrons are negatively charged, if there are more electrons than protons, the atom or molecule will be negatively charged. This is called an an ion , from the Greek á¼€vÎ¬ , meaning 'up'.
Horse murders	The Horse murders scandal was a form of insurance fraud in the United States in which expensive horses, many of them show jumpers, were insured against death, accident and then killed to collect the insurance money. It is not known how many horses were murdered between the mid 1970s and the mid-1990s, when a Federal Bureau of Investigation (FBI) investigation brought the horse killings to light, but the number is thought to be well over 50, and may have been as high as 100. In addition, in 1977, the heiress Helen Brach disappeared and was presumed by law enforcement agents to have been murdered by the perpetrators of these crimes, because she threatened to report their criminal activity to authorities; continuing investigations into Brach's death began to uncover the insurance fraud in the 1990s.
Hydrogen	Hydrogen is the chemical element with atomic number 1. It is represented by the symbol H. At standard temperature and pressure, Hydrogen is a colorless, odorless, nonmetallic, tasteless, highly flammable diatomic gas with the molecular formula H_2. With an atomic weight of 1.007 94 u, Hydrogen is the lightest element.
Acetabularia	Acetabularia is a genus of green algae, specifically of the Polyphysaceae family, Typically found in subtropical waters, Acetabularia is a single-cell organism, but gigantic in size and complex in form, making it an excellent model organism for studying cell biology. In form, the mature Acetabularia resembles the round leaves of a nasturtium, being 0.5 to 10 cm tall and having three anatomical parts: a bottom rhizoid that resembles a set of short roots; a long stalk in the middle; and a top umbrella of branches that may fuse into a cap. The single nucleus of Acetabularia is located in the rhizoid, and allows the cell to regenerate completely if its cap is removed.
Barr body	In those species (including humans) in which sex is determined by the presence of the Y or W chromosome rather than the diploidy of the X or Z, a Barr body is the inactive X chromosome in a female cell 2003), rendered inactive in a process called Lyonization. The Lyon hypothesis states that in cells with multiple X chromosomes, all but one are inactivated during mammalian embryogenesis (Lyon, 1961.) This happens early in embryonic development at random in mammals, (Brown, 1997) except in marsupials and in some extra-embryonic tissues of some placental mammals, in which the father's X chromosome is always deactivated (Lee, 2003.)
Acetylcholine	The chemical compound Acetylcholine is a neurotransmitter in both the peripheral nervous system (PNS) and central nervous system (CNS) in many organisms including humans. Acetylcholine is one of many neurotransmitters in the autonomic nervous system (ANS) and the only neurotransmitter used in the motor division of the somatic nervous system. (Sensory neurons use glutamate and various peptides at their synapses.)

Acid	An acid is traditionally considered any chemical compound that, when dissolved in water, gives a solution with a hydrogen ion activity greater than in pure water, i.e. a pH less than 7.0. That approximates the modern definition of Johannes Nicolaus Brønsted and Martin Lowry, who independently defined an acid as a compound which donates a hydrogen ion (H^+) to another compound (called a base.) Common examples include acetic acid and sulfuric acid (used in car batteries.)
Ammonia	Ammonia is a compound of nitrogen and hydrogen with the formula NH_3. It is normally encountered as a gas with a characteristic pungent odor. Ammonia contributes significantly to the nutritional needs of terrestrial organisms by serving as a precursor to foodstuffs and fertilizers.
Basal body	A Basal body is an organelle formed from a centriole, a short cylindrical array of microtubules. It is found at the base of a eukaryotic undulipodium (cilium or flagellum) and serves as a nucleation site for the growth of the axoneme microtubules. Centrioles, from which basal bodies are derived, act as anchoring sites for proteins that in turn anchor microtubules within centrosomes, one type of microtubule organizing center (MTOC.)
Chromosome	A Chromosome is an organized structure of DNA and protein that is found in cells. It is a single piece of coiled DNA containing many genes, regulatory elements and other nucleotide sequences. Chromosome s also contain DNA-bound proteins, which serve to package the DNA and control its functions.
Scale	In most biological nomenclature, a Scale is a small rigid plate that grows out of an animal's skin to provide protection. In lepidopteran species, scales are plates on the surface of the insect wing, and provide coloration. Scales are quite common and have evolved multiple times with varying structure and function.

Cholesterol

Cholesterol is a lipidic, waxy steroid found in the cell membranes and transported in the blood plasma of all animals. It is an essential component of mammalian cell membranes where it is required to establish proper membrane permeability and fluidity. Cholesterol is the principal sterol synthesized by animals, but small quantities are synthesized in other eukaryotes, such as plants and fungi.

Carbon

Carbon is the chemical element with symbol C and atomic number 6. As a member of group 14 on the periodic table, it is nonmetallic and tetravalent--making four electrons available to form covalent chemical bonds. There are three naturally occurring isotopes, with ^{12}C and ^{13}C being stable, while ^{14}C is radioactive, decaying with a half-life of about 5730 years.

Horse murders

The Horse murders scandal was a form of insurance fraud in the United States in which expensive horses, many of them show jumpers, were insured against death, accident and then killed to collect the insurance money. It is not known how many horses were murdered between the mid 1970s and the mid-1990s, when a Federal Bureau of Investigation (FBI) investigation brought the horse killings to light, but the number is thought to be well over 50, and may have been as high as 100. In addition, in 1977, the heiress Helen Brach disappeared and was presumed by law enforcement agents to have been murdered by the perpetrators of these crimes, because she threatened to report their criminal activity to authorities; continuing investigations into Brach's death began to uncover the insurance fraud in the 1990s.

Heterochromatin

Heterochromatin is a tightly packed form of DNA. Its major characteristic is that transcription is limited. As such, it is a means to control gene expression, through regulation of the transcription initiation.

Chromatin is found in two varieties: euchromatin and Heterochromatin.

High-density lipoprotein

High-density lipoprotein is one of the five major groups of lipoproteins (chylomicrons, VLDL, IDL, LDL, HDL) which enable lipids like cholesterol and triglycerides to be transported within the water based blood stream. In healthy individuals, about thirty percent of blood cholesterol is carried by HDL .

It is hypothesized that HDL can remove cholesterol from atheroma within arteries and transport it back to the liver for excretion or re-utilization--which is the main reason why HDL-bound cholesterol is sometimes called 'good cholesterol', or HDL-C. A high level of HDL-C seems to protect against cardiovascular diseases, and low HDL cholesterol levels (less than 40 mg/dL) increase the risk for heart disease.

Inorganic

Traditionally, inorganic compounds are considered to be of a mineral, not biological, origin. Complementarily, most organic compounds are traditionally viewed as being of biological origin. Over the past century, the precise classification of inorganic vs organic compounds has become less important to scientists, primarily because the majority of known compounds are synthetic and not of natural origin.

Low-density lipoprotein

Low-density lipoprotein is a type of lipoprotein that transports cholesterol and triglycerides from the liver to peripheral tissues. LDL is one of the five major groups of lipoproteins; these groups include chylomicrons, very Low-density lipoprotein (VLDL), intermediate-density lipoprotein (IDL), Low-density lipoprotein, and high-density lipoprotein (HDL), although some alternative organizational schemes have been proposed. Like all lipoproteins, LDL enables fats and cholesterol to move within the water-based solution of the blood stream.

Group

In naming cultivated plants, a Group is a formal classification category, under the International Code of Nomenclature for Cultivated Plants (ICNCP):

> ICNCP Art. 3.1: 'a formal category for assembing cultivars, individual plants or assemblages of plants on the basis of defined similarity' .

The term 'Group' (with a capital G) was introduced in the 2004 ICNCP, replacing the 'Cultivar-Group' of the 1995 ICNCP.

A Group is united by some common trait; for example there may be a Group of yellow-flowering cultivars, a Group of cultivars with variegated leaves, a Group of cultivars resistant to a particular disease, etc.

Isomer

Isomer is an element of transverse body articulation of the bilateral fossil animals of the Phylum Proarticulata from the Ediacaran period. This term has been proposed by Andrey Yu. Ivantsov, a Russian paleontologist from the Laboratory of the Precambrian organisms, Paleontological Institute, Russian Academy of Sciences.

Skeleton

In biology, a Skeleton is a rigid framework that provides protection and structure in many types of animal, particularly those of the phylum Chordata and of the superphylum Ecdysozoa. Exo Skeleton s are external, as is typical of many invertebrates; they enclose the soft tissues and organs of the body. Exo Skeleton s may undergo periodic moulting as the animal grows.

Carbohydrates

Carbohydrates or saccharides are the most abundant of the four major classes of biomolecules. They fill numerous roles in living things, such as the storage and transport of energy (eg: starch, glycogen) and structural components (eg: cellulose in plants and chitin.) Additionally, Carbohydrates and their derivatives play major roles in the working process of the immune system, fertilization, pathogenesis, blood clotting, and development.

Decomposers

Decomposers are organisms that consume dead or decaying organisms, and, in doing so, carry out the natural process of decomposition. Like herbivores and predators, Decomposers are heterotrophic, meaning that they use organic substrates to get their energy, carbon and nutrients for growth and development. Decomposers use deceased organisms and non-living organic compounds as their food source.

Dehydration

Dehydration is defined as excessive loss of body water. It is literally the removal of water from an object. In physiological terms, it entails a relative deficiency of water molecules in relation to other dissolved solutes.

Lipids

Lipids are a broad group of naturally-occurring molecules which includes fats, waxes, sterols, fat-soluble vitamins (such as vitamins A, D, E and K), monoglycerides, diglycerides, phospholipids, and others. The main biological functions of lipids include energy storage, as structural components of cell membranes, and as important signaling molecules.

lipids may be broadly defined as hydrophobic or amphiphilic small molecules; the amphiphilic nature of some lipids allows them to form structures such as vesicles, liposomes, or membranes in an aqueous environment.

Deoxyribose

Deoxyribose, also known as D-Deoxyribose and 2-Deoxyribose, is an aldopentose -- a monosaccharide containing five carbon atoms, and including an aldehyde functional group in its linear structure. It is a deoxy sugar derived from the pentose sugar ribose by the replacement of the hydroxyl group at the 2 position with hydrogen, leading to the net loss of an oxygen atom. Replacement of the hydroxyl group also shifts the conformation of the ring from C3'-endo to C2'-endo.

Disaccharide

A Disaccharide is the carbohydrate formed when two monosaccharides undergo a condensation reaction which involves the elimination of a small molecule, such as water, from the functional groups only. Like monosaccharides, Disaccharide s also dissolve in water, taste sweet and are called sugars.

	Disaccharide is one of the four chemical groupings of carbohydrates (monosaccharide, Disaccharide oligosaccharide, and polysaccharide.)
Fermentation	Fermentation is the process of deriving energy from the oxidation of organic compounds, such as carbohydrates, using an endogenous electron acceptor, which is usually an organic compound. This is in contrast to cellular respiration, where electrons are donated to an exogenous electron acceptor, such as oxygen, via an electron transport chain. Fermentation does not necessarily have to be carried out in an anaerobic environment.
Fructose	Fructose is a simple reducing sugar found in many foods and is one of the three important dietary monosaccharides along with glucose and galactose. Honey, tree fruits, berries, melons, and some root vegetables, such as beets, sweet potatoes, parsnips, and onions, contain Fructose, usually in combination with glucose in the form of sucrose. Fructose is also derived from the digestion of granulated table sugar (sucrose), a disaccharide consisting of glucose and Fructose.
Galactose	Galactose is a type of sugar which is less sweet than glucose. It is considered a nutritive sweetener because it has food energy. Its name comes from the Ancient Greek word for milk, γÎ¬λακτος .
Gene	A Gene is the basic unit of heredity in a living organism. All living things depend on Gene s. Gene s hold the information to build and maintain their cells and pass Gene tic traits to offspring.
Globular protein	Globular protein s comprising 'globe'-like proteins that are more or less soluble in aqueous solutions (where they form colloidal solutions.) This main characteristic helps distinguishing them from fibrous proteins (the other class), which are practically insoluble. The term globin can refer more specifically to proteins including the globin fold.
Glucose	Glucose, a monosaccharide also known as grape sugar, blood sugar is a very important carbohydrate in biology. The living cell uses it as a source of energy and metabolic intermediate. Glucose is one of the main products of photosynthesis and starts cellular respiration in both prokaryotes and eukaryotes
Monosaccharides	Monosaccharides are the most basic unit of carbohydrates. They are the simplest form of sugar and are usually colorless, water-soluble, crystalline solids. Some Monosaccharides have a sweet taste.
Spirogyra	Spirogyra is a genus of filamentous green algae of the order Zygnematales and there are more than 400 species of Spirogyra in the world. Spirogyra measures approximately 10 to 100μm in width and may stretch centimeters long. Spirogyra is unbranched with cylindrical cells connected end to end in long green filaments.
Cellulose	Cellulose is an organic compound with the formula $(C_6H_{10}O_5)$Template:Chem/dispAAA, a polysaccharide consisting of a linear chain of several hundred to over ten thousand β(1→4) linked D-glucose units. Cellulose is the structural component of the primary cell wall of green plants, many forms of algae and the oomycetes. Some species of bacteria secrete it to form biofilms.

Chitin

Chitin$_n$ is a long-chain polymer of a N-acetylglucosamine, a derivative of glucose, and is found in many places throughout the natural world. It is the main component of the cell walls of fungi, the exoskeletons of arthropods, such as crustaceans and insects, including ants, beetles and butterflies, the radula of mollusks and the beaks of cephalopods, including squid and octopuses. Chitin has also proven useful for several medical and industrial purposes.

Glycogen

Glycogen is the molecule which functions as the secondary short term energy storage in animal cells. It is made primarily by the liver and the muscles, but can also be made by Glycogen esis within the brain and stomach. Glycogen is the analogue of starch, a less branched glucose polymer in plants, and is commonly referred to as animal starch, having a similar structure to amylopectin.

Polysaccharides

Polysaccharides are polymeric carbohydrate structures, formed of repeating units (either mono- or di-saccharides) joined together by glycosidic bonds. These structures are often linear, but may contain various degrees of branching. Polysaccharides are often quite heterogeneous, containing slight modifications of the repeating unit.

Starch

Starch or amylum is a polysaccharide carbohydrate consisting of a large number of glucose units joined together by glycosidic bonds. Starch is produced by all green plants as an energy store and is a major food source for humans.

Pure Starch is a white, tasteless and odorless powder that is insoluble in cold water or alcohol.

Acid

An acid is traditionally considered any chemical compound that, when dissolved in water, gives a solution with a hydrogen ion activity greater than in pure water, i.e. a pH less than 7.0. That approximates the modern definition of Johannes Nicolaus Brønsted and Martin Lowry, who independently defined an acid as a compound which donates a hydrogen ion (H^+) to another compound (called a base.) Common examples include acetic acid and sulfuric acid (used in car batteries.)

CD36

CD36 is an integral membrane protein found on the surface of many cell types in vertebrate animals and is also known as FAT, SCARB3, GP88, glycoprotein IV and glycoprotein IIIb CD36 is a member of the class B scavenger receptor family of cell surface proteins. CD36 binds many ligands including collagen, thrombospondin, erythrocytes parasitized with Plasmodium falciparum, oxidized low density lipoprotein, native lipoproteins, oxidized phospholipids, and long-chain fatty acids.

Fatty acid

In chemistry, especially biochemistry, a Fatty acid is a carboxylic acid often with a long unbranched aliphatic tail (chain), which is either saturated or unsaturated. Carboxylic acids as short as butyric acid (4 carbon atoms) are considered to be Fatty acid s, whereas Fatty acid s derived from natural fats and oils may be assumed to have at least eight carbon atoms, caprylic acid (octanoic acid), for example. The most abundant natural Fatty acid s have an even number of carbon atoms because their biosynthesis involves acetyl-CoA, a coenzyme carrying a two-carbon-atom group

Fatty acids

Fatty acids are an important source of energy for many organisms. Excess glucose can be stored efficiently as fat. Triglycerides yield more than twice as much energy for the same mass as do carbohydrates or proteins.

Triglycerides

Â·) (more properly known as Â·), TAG or triacylglyceride) is a glyceride in which the glycerol is esterified with three fatty acids. It is the main constituent of vegetable oil and animal fats.

Triglycerides are formed from a single molecule of glycerol, combined with three fatty acids on each of the OH groups, and make up most of fats digested by humans.

Chromosome	A Chromosome is an organized structure of DNA and protein that is found in cells. It is a single piece of coiled DNA containing many genes, regulatory elements and other nucleotide sequences. Chromosome s also contain DNA-bound proteins, which serve to package the DNA and control its functions.
Hormone	Hormone s are chemicals released by cells that affect cells in other parts of the body. Only a small amount of Hormone is required to alter cell metabolism. It is essentially a chemical messenger that transports a signal from one cell to another.
Membrane	A Membrane is a layer of material which serves as a selective barrier between two phases and remains impermeable to specific particles, molecules, or substances when exposed to the action of a driving force. Some components are allowed passage by the Membrane into a permeate stream, whereas others are retained by it and accumulate in the retentate stream. Membrane s can be of various thickness, with homogeneous or heterogeneous structure.
Monounsaturated fats	In biochemistry and nutrition, Monounsaturated fats are fatty acids that have a single double bond in the fatty acid chain and all of the remainder of the carbon atoms in the chain are single-bonded. By contrast, polyunsaturated fatty acids have more than one double bond. Fatty acids are long-chained molecules having a methyl group at one end and a carboxylic acid group at the other end.
Phospholipids	Phospholipids are a class of lipids and are a major component of all cell membranes. Most Phospholipids contain a diglyceride, a phosphate group, and a simple organic molecule such as choline; one exception to this rule is sphingomyelin, which is derived from sphingosine instead of glycerol. They are a type of molecule.
Polyunsaturated	In nutrition, polyunsaturated fat is an abbreviation of polyunsaturated fatty acid. That is a fatty acid in which more than one double bond exists within the representative molecule. That is, the molecule has two or more points on its structure capable of supporting hydrogen atoms not currently part of the structure.
Polyunsaturated fat	In nutrition, Polyunsaturated fat is an abbreviation of Polyunsaturated fat ty acid. That is a fatty acid in which more than one double bond exists within the representative molecule. That is, the molecule has two or more points on its structure capable of supporting hydrogen atoms not currently part of the structure.
Steroid	A Steroid is a terpenoid lipid characterized by its sterane or Steroid nucleus: a carbon skeleton with four fused rings, generally arranged in a 6-6-6-5 fashion. Steroid s vary by the functional groups attached to these rings and the oxidation state of the rings. Hundreds of distinct Steroid s are found in plants, animals, and fungi.
Trans fat	Trans fat is the common name for a type of unsaturated fat with trans-isomer fatty acid(s.) Trans fat s may be monounsaturated or polyunsaturated but never saturated. Unsaturated fat is a fat molecule, containing one or more double bonds between the carbon atoms.
Helicobacter	Helicobacter is a genus of Gram-negative bacteria possessing a characteristic helix shape. They were initially considered to be members of the Campylobacter genus, but since 1989 they have been grouped in their own genus. Some species have been found living in the lining of the upper gastrointestinal tract, as well as the liver of mammals and some birds..

Helicobacter pylori

Helicobacter pylori is a Gram-negative, microaerophilic bacterium that inhabits various areas of the stomach and duodenum. It causes a chronic low-level inflammation of the stomach lining and is strongly linked to the development of duodenal and gastric ulcers and stomach cancer. Over 80% of individuals infected with the bacterium are asymptomatic.

MyoD

MyoD is a protein with a key role in regulating muscle differentiation. MyoD belongs to a family of proteins known as myogenic regulatory factors . These bHLH (basic helix loop helix) transcription factors act sequentially in myogenic differentiation.

Enzymes

Enzymes are biomolecules that catalyze (i.e., increase the rates of) chemical reactions. Nearly all known Enzymes are proteins. However, certain RNA molecules can be effective biocatalysts too.

Hemoglobin

Hemoglobin is the iron-containing oxygen-transport metalloprotein in the red blood cells of vertebrates, and the tissues of some invertebrates.

In mammals, the protein makes up about 97% of the red blood cell's dry content, and around 35% of the total content . Hemoglobin transports oxygen from the lungs or gills to the rest of the body where it releases the oxygen for cell use.

Myosin

Myosin s are a large family of motor proteins found in eukaryotic tissues. They are responsible for actin-based motility.

> 'The term Myosin was originally used to describe a group of similar, but nonidentical, ATPases found in striated and smooth muscle cells.' From Pollard and Korn, 1973

Following the discovery, by Pollard and Korn, of enzymes with Myosin like function in Acanthamoeba castellanii, a large number of divergent Myosin genes have been discovered throughout eukaryotes.

Amino acid

In chemistry, an Amino acid is a molecule containing both amine and carboxyl functional groups. These molecules are particularly important in biochemistry, where this term refers to alpha- Amino acid s with the general formula $H_2NCHRCOOH$, where R is an organic substituent. In the alpha Amino acid s, the amino and carboxylate groups are attached to the same carbon atom, which is called the α-carbon.

Peptide

Peptide s are short polymers formed from the linking, in a defined order, of α-amino acids. The link between one amino acid residue and the next is known as an amide bond or a Peptide bond.

Proteins are poly Peptide molecules .

Polypeptide

Proteins are polypeptide molecules (or consist of multiple polypeptide subunits.) The distinction is that peptides are short and polypeptide s/proteins are long. There are several different conventions to determine these, all of which have caveats and nuances.

Collagen

Collagen is the main protein of connective tissue in animals and the most abundant protein in mammals, making up about 25% to 35% of the whole-body protein content. It is naturally found exclusively in metazoa, including sponges. In muscle tissue it serves as a major component of endomysium.

Keratin

Keratin s are a family of fibrous structural proteins; tough and insoluble, they form the hard but un-mineralized structures found in reptiles, birds, amphibians, and mammals. They are rivaled as biological materials in toughness only by chitin.

	There are various types of Keratin s within a single animal.
Dentition	Dentition is the development of teeth and their arrangement in the mouth. All mammals except the monotremes, the xenarthrans, the pangolins, and the cetaceans have up to four distinct types of teeth, with a maximum number for each. These are the incisor (cutting), the canine, the premolar, and the molar (grinding.)
Nucleic acid	A Nucleic acid is a macromolecule composed of chains of monomeric nucleotides. In biochemistry these molecules carry genetic information or form structures within cells. The most common Nucleic acid s are deoxyribo Nucleic acid (D Nucleic acid) and ribo Nucleic acid (R Nucleic acid .)
Nucleotide	Nucleotide s are molecules that, when joined together, make up the structural units of RNA and DNA. Additionally, Nucleotide s play central roles in metabolism. In that capacity, they serve as sources of chemical energy (adenosine triphosphate and guanosine triphosphate), participate in cellular signaling (cyclic guanosine monophosphate and cyclic adenosine monophosphate), and are incorporated into important cofactors of enzymatic reactions (coenzyme A, flavin adenine di Nucleotide , flavin mono Nucleotide , and nicotinamide adenine di Nucleotide phosphate.) Figure 1: Structural elements of the most common Nucleotide s Figure 2: Ribose structure indicating numbering of carbon atoms A Nucleotide is composed of a nucleobase (nitrogenous base) and a five-carbon sugar (either ribose or 2'-deoxyribose), and one to three phosphate groups.
Shigella	Shigella is a genus of Gram-negative, non-spore forming rod-shaped bacteria closely related to Escherichia coli and Salmonella. The causative agent of human shigellosis, Shigella cause disease in primates, but not in other mammals. It is only naturally found in humans and apes.
Shigella dysenteriae	Shigella dysenteriae is a species of the rod-shaped bacterial genus Shigella. Shigella can cause shigellosis (bacillary dysentery.) Shigellae are Gram-negative, non-spore-forming, facultatively anaerobic, non-motile bacteria.
Cell	The Cell is the structural and functional unit of all known living organisms. It is the smallest unit of an organism that is classified as living, and is often called the building block of life. Some organisms, such as most bacteria, are unicellular (consist of a single Cell.)

Blood cell

A Blood cell is any cell of any type normally found in blood. In mammals, these fall into three general categories:

· Red Blood cell s - Erythrocytes

· White Blood cell s- Leukocytes

· Platelets - Thrombocytes red and white human Blood cell s as seen under a microscope using a blue slide stain
Together, these three kinds of Blood cell s sum up for a total 45% of blood tissue

.

Red blood cells

Red blood cells are the most common type of blood cell and the vertebrate body's principal means of delivering oxygen to the body tissues via the blood. They take up oxygen in the lungs or gills and release it while squeezing through the body's capillaries. The cells are filled with hemoglobin, a biomolecule that can bind to oxygen.

Microvillus

Microvilli (singular: Microvillus) are microscopic cellular membrane protrusions that increase the surface area of cells, and are involved in a wide variety of functions, including absorption, secretion, cellular adhesion, and mechanotransduction. Duodenum with brush border (microvilli)

Thousands of microvilli form a structure called the brush border that is found on the apical surface of some epithelial cells, such as the small intestinal enterocyte and the kidney proximal tubule.

Microvilli also occur in sensory cells of the inner ear (as stereocilia), in the cells of taste buds, and in olfactory receptor cells.

Escherichia

Escherichia is a genus of Gram-negative, non-spore forming, facultatively anaerobic, rod-shaped bacteria from the family Enterobacteriaceae. Inhabitants of the gastrointestinal tracts of warm-blooded animals, Escherichia species provide a portion of the microbially-derived vitamin K for their host.

While many Escherichia are harmless commensals, particular strains of some species are human pathogens, and are known as the most common cause of urinary tract infections, significant sources of gastrointestinal disease, ranging from simple diarrhea to dysentery-like conditions, as well as a wide-range of other pathogenic states.

Escherichia coli

Escherichia coli , is a Gram negative bacterium that is commonly found in the lower intestine of warm-blooded organisms . Most E. coli strains are harmless, but some, such as serotype O157:H7, can cause serious food poisoning in humans, and are occasionally responsible for costly product recalls. The harmless strains are part of the normal flora of the gut, and can benefit their hosts by producing vitamin K_2, or by preventing the establishment of pathogenic bacteria within the intestine.

Capsule

In botany a Capsule is a type of simple, dry fruit produced by many species of flowering plants. A Capsule is a dehiscent structure composed of two or more carpels, that, at maturity, split apart (dehisce) to release the seeds within. In some capsules, the split occurs between carpels, and in others each carpel splits open.In yet others, seeds are released through openings or pores that form in the Capsule.

Cell theory

Cell theory refers to the idea that cells are the basic unit of structure in every living thing. Development of this theory during the mid 1600s was made possible by advances in microscopy. This theory is one of the foundations of biology.

Cell wall

A Cell wall is a tough, flexible and sometimes fairly rigid layer that surrounds some types of cells. It is located outside the cell membrane and provides these cells with structural support and protection, and also acts as a filtering mechanism. A major function of the Cell wall is to act as a pressure vessel, preventing over-expansion when water enters the cell.

Cystic fibrosis

Cystic fibrosis is a genetic disorder known to be an inherited disease of the secretory glands, including the glands that make mucus and sweat.

The hallmarks of Cystic fibrosis are salty tasting skin, normal appetite but poor growth and poor weight gain, excess mucus production, and coughing/shortness of breath. Males can be infertile due to the condition congenital bilateral absence of the vas deferens.

Cytoplasm

The Cytoplasm is the part of a cell that is enclosed within the plasma membrane. In eukaryotic cells, the Cytoplasm contains organelles, such as mitochondria, which are filled with liquid that is kept separate from the rest of the Cytoplasm by biological membranes. The Cytoplasm is the site where most cellular activities occur, such as many metabolic pathways like glycolysis, and processes such as cell division.

Nucleoid

The Nucleoid is an irregularly-shaped region within the cell of prokaryotes which has nuclear material without a nuclear membrane and where the genetic material is localized. The genome of prokaryotic organisms generally is a circular, double-stranded piece of DNA, of which multiple copies may exist at any time. The length of a genome widely varies, but generally is at least a few million base pairs.

Binary fission

Binary fission is the form of asexual reproduction and cell division used by all prokaryotic and some eukaryotic organisms. This process results in the reproduction of a living prokaryotic cell by division into two parts which each have the potential to grow to the size of the original cell.

Mitosis and cytokinesis are not the same as Binary fission.

Fimbria

In bacteriology, Fimbria is a proteinaceous appendage in many gram-negative and gram positive bacteria that is thinner and shorter than a flagellum. This appendage ranges from 3-10 nanometers in diameter and can be up to several micrometers long. Fimbriae are used by bacteria to adhere to one another and to adhere to animal cells, and some inanimate objects.

Gene

A Gene is the basic unit of heredity in a living organism. All living things depend on Gene s. Gene s hold the information to build and maintain their cells and pass Gene tic traits to offspring.

Ribosome

Ribosome s are complexes of RNA and protein that are found in all cells with nuclei. Ribosome s from bacteria, archaea and eukaryotes (the three domains of life on Earth), have significantly different structure and RNA. The Ribosome s in the mitochondria of eukaryotic cells resemble those in bacteria, reflecting the evolutionary origin of this organelle.

The Ribosome functions in the expression of the genetic code from nucleic acid into protein, in a process called translation.

Chromosome	A Chromosome is an organized structure of DNA and protein that is found in cells. It is a single piece of coiled DNA containing many genes, regulatory elements and other nucleotide sequences. Chromosome s also contain DNA-bound proteins, which serve to package the DNA and control its functions.
Membrane	A Membrane is a layer of material which serves as a selective barrier between two phases and remains impermeable to specific particles, molecules, or substances when exposed to the action of a driving force. Some components are allowed passage by the Membrane into a permeate stream, whereas others are retained by it and accumulate in the retentate stream. Membrane s can be of various thickness, with homogeneous or heterogeneous structure.
Protein	Protein s are organic compounds made of amino acids arranged in a linear chain. The amino acids in a polymer chain are joined together by the peptide bonds between the carboxyl and amino groups of adjacent amino acid residues. The sequence of amino acids in a protein is defined by the sequence of a gene, which is encoded in the genetic code.
Receptor	In biochemistry, a Receptor is a protein molecule, embedded in either the plasma membrane or cytoplasm of a cell, to which a mobile signaling (or 'signal') molecule may attach. A molecule which binds to a Receptor is called a 'ligand,' and may be a peptide (such as a neurotransmitter), a hormone, a pharmaceutical drug, or a toxin, and when such binding occurs, the Receptor undergoes a conformational change which ordinarily initiates a cellular response. However, some ligands merely block Receptor s without inducing any response (e.g. antagonists.)
Organelle	In cell biology, an Organelle is a specialized subunit within a cell that has a specific function, and is usually separately enclosed within its own lipid membrane. A typical animal cell. Within the cytoplasm, the major Organelle s and cellular structures include: (1) nucleolus (2) nucleus (3) ribosome (4) vesicle (5) rough endoplasmic reticulum (6) Golgi apparatus (7) cytoskeleton (8) smooth endoplasmic reticulum (9) mitochondria (10) vacuole (11) cytosol (12) lysosome (13) centriole. The name Organelle comes from the idea that these structures are to cells what an organ is to the body (hence the name Organelle the suffix -elle being a diminutive.)
Cytoskeleton	The Cytoskeleton is a cellular 'scaffolding' or 'skeleton' contained within the cytoplasm. The Cytoskeleton is present in all cells; it was once thought this structure was unique to eukaryotes, but recent research has identified the prokaryotic Cytoskeleton. It is a dynamic structure that maintains cell shape, protects the cell, enables cellular motion (using structures such as flagella, cilia and lamellipodia), and plays important roles in both intracellular transport (the movement of vesicles and organelles, for example) and cellular division.
Vesicle	A Vesicle is a small bubble of liquid within a cell. More technically, a Vesicle is a small, intracellular, membrane-enclosed sac that stores or transports substances within a cell. Vesicles form naturally because of the properties of lipid membranes
RNA	Ribonucleic acid (RNA) is a biologically important type of molecule that consists of a long chain of nucleotide units. Each nucleotide consists of a nitrogenous base, a ribose sugar, and a phosphate. RNA is very similar to DNA, but differs in a few important structural details: in the cell, RNA is usually single-stranded, while DNA is usually double-stranded; RNA nucleotides contain ribose while DNA contains deoxyribose (a type of ribose that lacks one oxygen atom); and RNA has the base uracil rather than thymine that is present in DNA.

RNA is transcribed from DNA by enzymes called RNA polymerases and is generally further processed by other enzymes.

Chromatin

Chromatin is the complex combination of DNA, RNA, and protein that makes up chromosomes. It is found inside the nuclei of eukaryotic cells, and within the nucleoid in prokaryotic cells. It is divided between hetero Chromatin (condensed) and eu Chromatin (extended) forms.

Messenger ribonucleic acid

Messenger ribonucleic acid is a molecule of RNA encoding a chemical 'blueprint' for a protein product. mRNA is transcribed from a DNA template, and carries coding information to the sites of protein synthesis: the ribosomes. Here, the nucleic acid polymer is translated into a polymer of amino acids: a protein.

Nucleolus

The Nucleolus is a non-membrane bound structure composed of protein and nucleic acids found within the nucleus. The ribosomal RNA is transcribed within the Nucleolus. The Nucleolus ultrastructure can be visualized through an electron microscope, while the organization and dynamics can be studied through fluorescent protein tagging and fluorescent recovery after photobleaching

Ribosomal RNA

Ribosomal RNA is the central component of the ribosome, the protein manufacturing machinery of all living cells. The function of the rRNA is to provide a mechanism for decoding mRNA into amino acids and to interact with the tRNAs during translation by providing peptidyl transferase activity.The tRNA then brings the necessary amino acids corresponding to the appropriate mRNA codon.

The ribosome is composed of two subunits P, and E.

· The A site in the ribosome binds to an aminoacyl-tRNA (a tRNA bound to an amino acid.)
· The amino (NH2) group of the aminoacyl-tRNA, which contains the new amino acid, attacks the ester linkage of peptidyl-tRNA (contained within the P site), which contains the last amino acid of the growing chain, forming a new peptide bond. This reaction is catalyzed by peptidyl transferase.
· The tRNA that was holding on the last amino acid is moved to the E site, and what used to be the aminoacyl-tRNA is now the peptidyl-tRNA.

A single mRNA can be translated simultaneously by multiple ribosomes.

Both prokaryotic and eukaryotic can be broken down into two subunits (the S in 16S represents Svedberg units):

Note that the S units of the subunits cannot simply be added because they represent measures of sedimentation rate rather than of mass.

Endoplasmic reticulum

The Endoplasmic reticulum is a eukaryotic organelle that forms an interconnected network of tubules, vesicles, and cisternae within cells. The lacey membranes of the Endoplasmic reticulum were first seen by Keith R. Porter, Albert Claude, and Ernest F. Fullam in 1945.

These structures are responsible for several specialized functions: protein translation, folding and transport of proteins to be used in the cell membrane (e.g. transmembrane receptors and other integral membrane proteins), or to be secreted (exocytosed) from the cell (e.g. digestive enzymes); sequestration of calcium; and production and storage of glycogen, steroids, and other macromolecules.

Nuclear envelope	The Nuclear envelope is a double lipid bilayer that encloses the genetic material in eukaryotic cells. The Nuclear envelope also serves as the physical barrier, separating the contents of the nucleus from the cytosol Many nuclear pores are inserted in the Nuclear envelope, which facilitate and regulate the exchange of materials between the nucleus and the cytoplasm.
Nuclear pore	Nuclear pore s are large protein complexes that cross the nuclear envelope, which is the double membrane surrounding the eukaryotic cell nucleus. There are about on average 2000 Nuclear pore complexes in the nuclear envelope of a vertebrate cell, but it varies depending on cell type and the stage in the life cycle. The proteins that make up the Nuclear pore complex are known as nucleoporins.
Polyribosomes	Polyribosomes are a cluster of ribosomes, bound to a mRNA molecule, first discovered and characterized by Jonathan Warner, Paul Knopf, and Alex Rich in 1963. Polyribosomes read one strand of mRNA simultaneously, helping to synthesize the same protein at different spots on the mRNA, mRNA being the 'messenger' in the process of protein synthesis. They may appear as clusters, linear arrays, or rosettes in routine: this is aided by the fact that mRNA is able to be twisted into a circular formation, creating a cycle of rapid ribosome recycling, and utilization of ribosomes.
Golgi apparatus	The Golgi apparatus is an organelle found in most eukaryotic cells. It was identified in 1898 by the Italian physician Camillo Golgi and was named after him. The primary function of the Golgi apparatus is to process and package macromolecules, such as proteins and lipids, after their synthesis and before they make their way to their destination; it is particularly important in the processing of proteins for secretion.
Shigella	Shigella is a genus of Gram-negative, non-spore forming rod-shaped bacteria closely related to Escherichia coli and Salmonella. The causative agent of human shigellosis, Shigella cause disease in primates, but not in other mammals. It is only naturally found in humans and apes.
Shigella dysenteriae	Shigella dysenteriae is a species of the rod-shaped bacterial genus Shigella. Shigella can cause shigellosis (bacillary dysentery.) Shigellae are Gram-negative, non-spore-forming, facultatively anaerobic, non-motile bacteria.
Horse murders	The Horse murders scandal was a form of insurance fraud in the United States in which expensive horses, many of them show jumpers, were insured against death, accident and then killed to collect the insurance money. It is not known how many horses were murdered between the mid 1970s and the mid-1990s, when a Federal Bureau of Investigation (FBI) investigation brought the horse killings to light, but the number is thought to be well over 50, and may have been as high as 100. In addition, in 1977, the heiress Helen Brach disappeared and was presumed by law enforcement agents to have been murdered by the perpetrators of these crimes, because she threatened to report their criminal activity to authorities; continuing investigations into Brach's death began to uncover the insurance fraud in the 1990s.
Endomembrane system	The Endomembrane system is composed of the different membranes that are suspended in the cytoplasm within a eukaryotic cell. These membranes divide the cell into functional and structural compartments, or organelles. In eukaryotes the organelles of the Endomembrane system include: the nuclear envelope, the endoplasmic reticulum, the Golgi apparatus, lysosomes, vacuoles, vesicles, and the cell membrane.

Lysosomes	Lysosomes are organelles containing digestive enzymes (acid hydrolases.) They are found in animal cells, while in plant cells the same roles are performed by the vacuole. They digest excess or worn-out organelles, food particles, and engulfed viruses or bacteria.
Secretion	Secretion is the process of elaborating and releasing chemicals from a cell, a secreted chemical substance or amount of substance. In contrast to excretion, the substance may have a certain function, rather than being a waste product. Secretion in bacterial species means the transport or translocation of effector molecules for example proteins, enzymes or toxins (such as cholera toxin in pathogenic bacteria for example Vibrio cholerae) from across the interior (cytoplasm or cytosol) of a bacterial cell to its exterior.
Chlamydomonas	Chlamydomonas is a genus of green alga. They are unicellular flagellates. Chlamydomonas is used as a model organism for molecular biology, especially studies of flagellar motility and chloroplast dynamics, biogenesis, and genetics.
Adenosine	Adenosine is a nucleoside composed of a molecule of adenine attached to a ribose sugar molecule (ribofuranose) moiety via a β-N_9-glycosidic bond. Adenosine plays an important role in biochemical processes, such as energy transfer--as Adenosine triphosphate (ATP) and Adenosine diphosphate (ADP)--as well as in signal transduction as cyclic Adenosine monophosphate, cAMP. It is also an inhibitory neurotransmitter, believed to play a role in promoting sleep and suppressing arousal, with levels increasing with each hour an organism is awake. Adenosine is an endogenous purine nucleoside that modulates many physiological processes.
Deoxyadenosine diphosphate	Deoxyadenosine diphosphate is a derivative of the common nucleic acid ATP in which the -OH (hydroxyl) group on the 2' carbon on the nucleotide's pentose has been removed (hence the deoxy- part of the name.) Additionally, the diphosphate of the name indicates that one of the phosphoryl groups of ATP has been removed, most likely by hydrolysis. Deoxyadenosine diphosphate would be abbreviated dADP.
Chloroplast	Chloroplast s are organelles found in plant cells and other eukaryotic organisms that conduct photosynthesis. Chloroplast s capture light energy to conserve free energy in the form of ATP and reduce NADP to NADPH through a complex set of processes called photosynthesis. The word Chloroplast is derived from the Greek words chloros which means green and plast which means form or entity.
Thylakoid	A Thylakoid is a membrane-bound compartment inside chloroplasts and cyanobacteria. They are the site of the light-dependent reactions of photosynthesis. The word 'Thylakoid' is derived from the Greek thylakos, meaning 'sac'.
Cellular respiration	Cellular respiration is the set of the metabolic reactions and processes that take place in organisms' cells to convert biochemical energy from nutrients into adenosine triphosphate (ATP), and then release waste products. The reactions involved in respiration are catabolic reactions that involve the oxidation of one molecule and the reduction of another. Nutrients commonly used by animal and plant cells in respiration include glucose, amino acids and fatty acids, and a common oxidizing agent (electron acceptor) is molecular oxygen (O_2.)

Cristae

Cristae are the internal compartments formed by the inner membrane of a mitochondrion. They are studded with proteins, including ATP synthase and a variety of cytochromes. The maximum surface for chemical reactions to occur is within the mitochondria.

Pigment

A pigment is the material that changes the color of light it reflects as the result of selective color absorption. This physical process differs from fluorescence, phosphorescence, and other forms of luminescence, in which the material itself emits light.

Many materials selectively absorb certain wavelengths of light.

Actin

Actin is a globular, roughly 42-kDa highly conserved protein found in all eukaryotic cells (the only known exception being nematode sperm) where it may be present at concentrations of over 100 μM. It is also one of the most highly-conserved proteins, differing by no more than 20% in species as diverse as algae and humans. Actin is the monomeric subunit of two types of filaments in cells: microfilaments, one of the three major components of the cytoskeleton, and thin filaments, part of the contractile apparatus in muscle cells. Thus, Actin participates in many important cellular processes including muscle contraction, cell motility, cell division and cytokinesis, vesicle and organelle movement, cell signaling, and the establishment and maintenance of cell junctions and cell shape.

Microfilaments

Microfilaments are the thinnest filaments of the cytoskeleton found in the cytoplasm of all eukaryotic cells. These linear polymers of actin subunits are flexible and relatively strong, resisting buckling by multi-piconewton compressive forces and filament fracture by nanonewton tensile forces. Microfilaments are highly versatile, functioning in (a) actoclampin-driven expansile molecular motors, where each elongating filament harnesses the hydrolysis energy of its 'on-board' ATP to drive actoclampin end-tracking motors to propel cell crawling, ameboid movement, and changes in cell shape, and (b) actomyosin-driven contractile molecular motors, where the thin filaments serve as tensile platforms for myosin's ATP hydrolysis-dependent pulling action in muscle contraction and uropod advancement.

Centrosome

In cell biology, the Centrosome is an organelle that serves as the main microtubule organizing center (MTOC) of the animal cell as well as a regulator of cell-cycle progression. It was discovered by Edouard Van Beneden in 1883 and was described and named in 1888 by Theodor Boveri. The Centrosome is thought to have evolved only in the metazoan lineage of eukaryotic cells.

Dynein

Dynein is a motor protein in cells which converts the chemical energy contained in ATP into the mechanical energy of movement. Dynein transports various cellular cargo by 'walking' along cytoskeletal microtubules towards the minus-end of the microtubule, which is usually oriented towards the cell center. Thus, they are called 'minus-end directed motors,' while kinesins, motor proteins that move toward the microtubules' plus end, are called plus-end directed motors.

Intermediate filament

Intermediate filament s (Intermediate filament s) are a family of related proteins that share common structural and sequence features. Intermediate filament s have an average diameter of 10 nanometers, which is between that of actin (microfilaments) and microtubules, although they were initially designated 'intermediate' because their average diameter was between those of narrower microtubules and wider myosin filaments. Most types of Intermediate filament s are cytoplasmic, but one type, the lamins, are nuclear.

Kinesin	Kinesin s are a class of motor proteins found in eukaryotic cells. Kinesin s move along microtubule cables powered by the dephosphorylation of ATP (thus Kinesin s are ATPases.) The active movement of Kinesin s supports several cellular functions including mitosis, meiosis and transport of cargo such as axonal transport.
Microtubules	Microtubules are one of the components of the cytoskeleton. They have a diameter of 25 nm and length varying from 200 nanometers to 25 micrometers. Microtubules serve as structural components within cells and are involved in many cellular processes including mitosis, cytokinesis, and vesicular transport.
Motor protein	Motor protein s are a class of molecular motors that are able to move along the surface of a suitable substrate. They are powered by the hydrolysis of ATP and convert chemical energy into mechanical work. The most prominent example of a Motor protein is the muscle protein myosin which 'motors' the contraction of muscle fibers in animals.
Centriole	A Centriole is a barrel-shaped organelle found in most animal eukaryotic cells, though absent in higher plants and most fungi. The walls of each Centriole are usually composed of nine triplets of microtubules (protein of the cytoskeleton.) Deviations from this structure include Drosophila melanogaster embryos, with nine doublets, and Caenorhabditis elegans sperm cells and early embryos, with nine singlets.
Extracellular	In cell biology, molecular biology and related fields, the word Extracellular means 'outside the cell'. This space is usually taken to be outside the plasma membranes, and occupied by fluid. The term is used in contrast to intracellular (inside the cell.)
Plasmodesmata	Plasmodesmata are microscopic channels which traverse the cell walls of plant cells and some algal cells enabling transport and communication between them. Species that have Plasmodesmata include members of the Charophyceae, Charales and Coleochaetales (which are all algae), as well as all embryophytes, better known as land plants. Unlike animal cells, every plant cell is surrounded by a polysaccharide cell wall.
Gap junction	A Gap junction or nexus is a specialized intercellular connection between certain animal cell-types. It directly connects the cytoplasm of two cells, which allows various molecules and ions to pass freely between cells. One Gap junction is composed of two connexons (or hemichannels) which connect across the intercellular space.
Tight junctions	Tight junctions are the closely associated areas of two cells whose membranes join together forming a virtually impermeable barrier to fluid. It is a type of junctional complex present only in vertebrates. The corresponding junctions that occur in invertebrates are septate junctions.

Adenosine

Adenosine is a nucleoside composed of a molecule of adenine attached to a ribose sugar molecule (ribofuranose) moiety via a β-N_9-glycosidic bond.

Adenosine plays an important role in biochemical processes, such as energy transfer--as Adenosine triphosphate (ATP) and Adenosine diphosphate (ADP)--as well as in signal transduction as cyclic Adenosine monophosphate, cAMP. It is also an inhibitory neurotransmitter, believed to play a role in promoting sleep and suppressing arousal, with levels increasing with each hour an organism is awake.

Adenosine is an endogenous purine nucleoside that modulates many physiological processes.

Deoxyadenosine diphosphate

Deoxyadenosine diphosphate is a derivative of the common nucleic acid ATP in which the -OH (hydroxyl) group on the 2' carbon on the nucleotide's pentose has been removed (hence the deoxy- part of the name.) Additionally, the diphosphate of the name indicates that one of the phosphoryl groups of ATP has been removed, most likely by hydrolysis.

Deoxyadenosine diphosphate would be abbreviated dADP.

Horse murders

The Horse murders scandal was a form of insurance fraud in the United States in which expensive horses, many of them show jumpers, were insured against death, accident and then killed to collect the insurance money. It is not known how many horses were murdered between the mid 1970s and the mid-1990s, when a Federal Bureau of Investigation (FBI) investigation brought the horse killings to light, but the number is thought to be well over 50, and may have been as high as 100. In addition, in 1977, the heiress Helen Brach disappeared and was presumed by law enforcement agents to have been murdered by the perpetrators of these crimes, because she threatened to report their criminal activity to authorities; continuing investigations into Brach's death began to uncover the insurance fraud in the 1990s.

Leukemia inhibitory factor

Leukemia inhibitory factor an interleukin 6 class cytokine, is a chemical in cells that affects their growth and development.

Leukemia inhibitory factor derives its name from its ability to induce the terminal differentiation of myeloid leukaemic cells. Other properties attributed to the cytokine include: the growth promotion and cell differentiation of different types of target cells, influence on bone metabolism, cachexia, neural development, embryogenesis and inflammation.

Active site

The Active site of an enzyme contains the catalytic and binding sites. The structure and chemical properties of the Active site allow the recognition and binding of the substrate.

The Active site is usually a big pocket or cleft surrounded by amino acid- and other side chains at the surface of the enzyme that contains residues responsible for the substrate specificity (charge, hydrophobicity, steric hindrance) and catalytic residues which often act as proton donors or acceptors or are responsible for binding a cofactor such as PLP, TPP or NAD. The Active site is also the site of inhibition of enzymes

Enzymes

Enzymes are biomolecules that catalyze (i.e., increase the rates of) chemical reactions. Nearly all known Enzymes are proteins. However, certain RNA molecules can be effective biocatalysts too.

Metabolic pathway

In biochemistry, a Metabolic pathway is a series of chemical reactions occurring within a cell. In each pathway, a principal chemical is modified by chemical reactions. Enzymes catalyze these reactions, and often require dietary minerals, vitamins, and other cofactors in order to function properly.

Substrate
In biochemistry, a substrate is a molecule upon which an enzyme acts. Enzymes catalyze chemical reactions involving the substrate(s.) In the case of a single substrate, the substrate binds with the enzyme active site, and an enzyme-substrate complex is formed.

APC
APC (adenomatosis polyposis coli) is a human gene that is classified as a tumor suppressor gene. Tumor suppressor genes prevent the uncontrolled growth of cells that may result in cancerous tumors. The protein made by the APC gene plays a critical role in several cellular processes that determine whether a cell may develop into a tumor.

Shigella
Shigella is a genus of Gram-negative, non-spore forming rod-shaped bacteria closely related to Escherichia coli and Salmonella. The causative agent of human shigellosis, Shigella cause disease in primates, but not in other mammals. It is only naturally found in humans and apes.

Shigella dysenteriae
Shigella dysenteriae is a species of the rod-shaped bacterial genus Shigella. Shigella can cause shigellosis (bacillary dysentery.) Shigellae are Gram-negative, non-spore-forming, facultatively anaerobic, non-motile bacteria.

Cell
The Cell is the structural and functional unit of all known living organisms. It is the smallest unit of an organism that is classified as living, and is often called the building block of life. Some organisms, such as most bacteria, are unicellular (consist of a single Cell.)

Diffusion
Molecular diffusion, often called simply diffusion, is a net transport of molecules from a region of higher concentration to one of lower concentration by random molecular motion. The result of diffusion is a gradual mixing of material. In a phase with uniform temperature, absent external net forces acting on the particles, the diffusion process will eventually result in complete mixing or a state of equilibrium.

Passive transport
Passive transport means moving biochemicals and atomic or molecular substances across the cell membrane. Unlike active transport, this process does not involve chemical energy. The four main kinds of Passive transport are diffusion, facilitated diffusion, filtration and osmosis.

Active transport
Active transport is the mediated process of moving particles across a biological membrane against a concentration gradient. If the process uses chemical energy, such as from adenosine triphosphate (Active transport P), it is termed primary Active transport Secondary Active transport involves the use of an electrochemical gradient.

Plasmolysis
Plasmolysis is the process in plant cells where the plasma membrane pulls away from the cell wall due to the loss of water through osmosis. The reverse process, de Plasmolysis , can occur if the cell is in a hypotonic solution resulting in a higher external osmotic pressure and net flow of water into the cell. Through observation of Plasmolysis and de Plasmolysis it is possible to determine the tonicity of the cell's environment as well as the rate solute molecules cross the cellular membrane.

Endocytosis
Endocytosis is the process by which cells absorb material (molecules such as proteins) from outside the cell by engulfing it with their cell membrane. It is used by all cells of the body because most substances important to them are large polar molecules that cannot pass through the hydrophobic plasma membrane or cell membrane. The process opposite to Endocytosis is exocytosis.

Exocytosis	Exocytosis is the durable process by which a cell directs the contents of secretory vesicles out of the cell membrane. These membrane-bound vesicles contain soluble proteins to be secreted to the extracellular environment, as well as membrane proteins and lipids that are sent to become components of the cell membrane. In multicellular organisms there are two types of Exocytosis: 1) Ca^{2+} triggered non-constitutive and 2) non Ca^{2+} triggered constitutive.
Mode of action	Historically, pesticides have often been classified according to their chemical groups and this is useful for understanding the properties of a given compound. However, it is the mode of action group which possibly represents the most useful pesticide classification for biologists. For example, MoA at the top of entries in the Pesticide Manual may be something like: 'FRAC G1', 'IRAC 2A' or HRAC G' - what do these mean? From a pesticide industry point of view, one of the most important threats to product sustainability and innovation is the onset of resistance.
Pinocytosis	In cellular biology, Pinocytosis is a form of endocytosis in which small particles are brought into the cell suspended within small vesicles which subsequently fuse with lysosomes to hydrolyze the particles. This process requires energy in the form of adenosine triphosphate (ATP), the chemical compound used as energy in a majority of cells. Pinocytosis is primarily used for the absorption of extracellular fluids (ECF), and in contrast to phagocytosis, generates very small vesicles.
Receptor-mediated endocytosis	Receptor-mediated endocytosis is a process by which cells internalize molecules by the inward budding of plasma membrane vesicles containing proteins with receptor sites specific to the molecules being internalized. After the binding of a ligand to plasma membrane spanning receptors, a signal is sent through the membrane, leading to membrane coating, and formation of a membrane invagination. The receptor, its ligand, and anything nearby are then internalised in sub-micrometre sized clathrin-coated vesicles.

Chromosome

A Chromosome is an organized structure of DNA and protein that is found in cells. It is a single piece of coiled DNA containing many genes, regulatory elements and other nucleotide sequences. Chromosome s also contain DNA-bound proteins, which serve to package the DNA and control its functions.

Photosynthesis

Photosynthesis is a process that converts carbon dioxide into organic compounds, especially sugars, using the energy from sunlight. Photosynthesis occurs in plants, algae, and many species of Bacteria, but not in Archaea. Photosynthetic organisms are called photoautotrophs, since it allows them to create their own food.

Chlamydomonas

Chlamydomonas is a genus of green alga. They are unicellular flagellates. Chlamydomonas is used as a model organism for molecular biology, especially studies of flagellar motility and chloroplast dynamics, biogenesis, and genetics.

Anatomy

The anatomy of spiders is in some aspects similar, but also different from that of other arthropods. The following characteristics are common to all spiders: A body with two segments, eight legs, spinnerets, no chewing parts, no wings, and the presence of chelicerae, which spiders use to hold prey, and in most cases, inject venom. Spiders have non-compound eyes, with most species having eight; the spiders known as Haplogynae may have six or fewer, and certain cave-dwelling spiders may have none at all.

Chlorophyll

Chlorophyll is a green pigment found in most plants, algae, and cyanobacteria. Chlorophyll absorbs light most strongly in the blue and red but poorly in the green portions of the electromagnetic spectrum, hence the green colour of Chlorophyll-containing tissues such as plant leaves.

Chlorophyll is vital for photosynthesis, which allows plants to obtain energy from light.

Chloroplast

Chloroplast s are organelles found in plant cells and other eukaryotic organisms that conduct photosynthesis. Chloroplast s capture light energy to conserve free energy in the form of ATP and reduce NADP to NADPH through a complex set of processes called photosynthesis.

The word Chloroplast is derived from the Greek words chloros which means green and plast which means form or entity.

Thylakoid

A Thylakoid is a membrane-bound compartment inside chloroplasts and cyanobacteria. They are the site of the light-dependent reactions of photosynthesis. The word 'Thylakoid' is derived from the Greek thylakos, meaning 'sac'.

Calvin cycle

The Calvin cycle is a series of biochemical reactions that take place in the stroma of chloroplasts in photosynthetic organisms. It was discovered by Melvin Calvin, James Bassham and Andrew Benson at the University of California, Berkeley . It is one of the light-independent reactions or dark reactions.

Light-dependent reactions

The Light-dependent reactions are the first stage of photosynthesis. In this process light energy is converted into chemical energy, in the form of the energy-carriers ATP and NADPH. In the light-independent reactions, the formed NADPH and ATP drive the reduction of CO_2 to more useful organic compounds, such as glucose.

The Light-dependent reactions take place on the thylakoid membrane inside a chloroplast.

Punnett square	The Punnett square is a diagram that is used to predict the outcome of a particular cross or breeding experiment. It is named after Reginald C. Punnett, who devised the approach, and is used by biologists to determine the probability of an offspring having a particular genotype. The Punnett square is a summary of every possible combination of one maternal allele with one paternal allele for each gene being studied in the cross.
Carotenoids	Carotenoids are organic pigments that are naturally occurring in the chloroplasts and chromoplasts of plants and some other photosynthetic organisms like algae, some types of fungus and some bacteria. There are over 600 known Carotenoids; they are split into two classes, xanthophylls (which contain oxygen) and carotenes (which are purely hydrocarbons, and contain no oxygen.) Carotenoids in general absorb blue light.
Edward	The Edward mango is a monoembryonic mango cultivar grown predominantly in Florida. It is considered by many to be among the finest tasting mangoes in the world; however, its poor yields have restrained the Edward from developing into a commercially significant variety. The Edward was first propagated in the 1920s by Edward Simmonds of the Plant Introduction Garden in Miami, Florida and is believed to be a hybrid cross of Haden and Carabao mango cultivars.
NADPH	Nicotinamide adenine dinucleotide phosphate ($NADP^+$, in older notation triphosphopyridine nucleotide, TPN) is used in anabolic reactions, such as lipid and nucleic acid synthesis, which require NADPH as a reducing agent. NADPH is the reduced form of $NADP^+$, and $NADP^+$ is the oxidized form of NADPH. NADP+ differs from NAD+ by the presence in NADP+ of an additional phosphate group on the 2' position of the ribose ring that carries the adenine moiety. In chloroplasts, NADP is reduced by ferredoxin-NADP+ reductase in the last step of the electron chain of the light reactions of photosynthesis.
Electron transport chain	An Electron transport chain couples a chemical reaction between an electron donor (such as NADH) and an electron acceptor (such as O_2) to the transfer of H^+ ions across a membrane, through a set of mediating biochemical reactions. These H^+ ions are used to produce adenosine triphosphate (ATP), the main energy intermediate in living organisms, as they move back across the membrane. Electron transport chain s are used for extracting energy from sunlight (photosynthesis) and from redox reactions such as the oxidation of sugars (respiration.)
Photosystem	Photosystem s are protein complexes involved in photosynthesis. They are found in the thylakoid membranes of plants, algae and cyanobacteria , or in the cytoplasmic membrane of photosynthetic bacteria. A Photosystem (or Reaction Center) is an enzyme which uses light to reduce molecules.
ATP synthase	An ATP synthase is a general term for an enzyme that can synthesize adenosine triphosphate (ATP) from adenosine diphosphate (ADP) and inorganic phosphate by using some form of energy. This energy is often in the form of protons moving down an electrochemical gradient, such as from the lumen into the stroma of chloroplasts or from the inter-membrane space into the matrix in mitochondria. The overall reaction sequence is: $ADP + P_i \rightarrow ATP$ These enzymes are of crucial importance in almost all organisms, because ATP is the common 'energy currency' of cells.

Horse murders	The Horse murders scandal was a form of insurance fraud in the United States in which expensive horses, many of them show jumpers, were insured against death, accident and then killed to collect the insurance money. It is not known how many horses were murdered between the mid 1970s and the mid-1990s, when a Federal Bureau of Investigation (FBI) investigation brought the horse killings to light, but the number is thought to be well over 50, and may have been as high as 100. In addition, in 1977, the heiress Helen Brach disappeared and was presumed by law enforcement agents to have been murdered by the perpetrators of these crimes, because she threatened to report their criminal activity to authorities; continuing investigations into Brach's death began to uncover the insurance fraud in the 1990s.
Melvin Ellis Calvin	Melvin Ellis Calvin was an American chemist most famed for discovering the Calvin cycle along with Andrew Benson and James Bassham, for which he was awarded the 1961 Nobel Prize in Chemistry. He spent most of his five-decade career at the University of California, Berkeley. Calvin was born in St. Paul, Minnesota, the son of Jewish immigrants.
Carbon	Carbon is the chemical element with symbol C and atomic number 6. As a member of group 14 on the periodic table, it is nonmetallic and tetravalent--making four electrons available to form covalent chemical bonds. There are three naturally occurring isotopes, with ^{12}C and ^{13}C being stable, while ^{14}C is radioactive, decaying with a half-life of about 5730 years.
CD36	CD36 is an integral membrane protein found on the surface of many cell types in vertebrate animals and is also known as FAT, SCARB3, GP88, glycoprotein IV and glycoprotein IIIb CD36 is a member of the class B scavenger receptor family of cell surface proteins. CD36 binds many ligands including collagen, thrombospondin, erythrocytes parasitized with Plasmodium falciparum, oxidized low density lipoprotein, native lipoproteins, oxidized phospholipids, and long-chain fatty acids.
Plants	Plants are living organisms belonging to the kingdom Plantae. They include familiar organisms such as trees, herbs, bushes, grasses, vines, ferns, mosses, and green algae. About 350,000 species of Plants, defined as seed Plants, bryophytes, ferns and fern allies, are estimated to exist currently.
Crassulacean acid metabolism	Crassulacean acid metabolism is an elaborate carbon fixation pathway in some plants. These plants fix carbon dioxide during the night, storing it as the four carbon acid malate. The CO_2 is released during the day, where it is concentrated around the enzyme RuBisCO, increasing the efficiency of photosynthesis.

Cell

The Cell is the structural and functional unit of all known living organisms. It is the smallest unit of an organism that is classified as living, and is often called the building block of life. Some organisms, such as most bacteria, are unicellular (consist of a single Cell.)

Cellular respiration

Cellular respiration is the set of the metabolic reactions and processes that take place in organisms' cells to convert biochemical energy from nutrients into adenosine triphosphate (ATP), and then release waste products. The reactions involved in respiration are catabolic reactions that involve the oxidation of one molecule and the reduction of another.

Nutrients commonly used by animal and plant cells in respiration include glucose, amino acids and fatty acids, and a common oxidizing agent (electron acceptor) is molecular oxygen (O_2.)

Edward

The Edward mango is a monoembryonic mango cultivar grown predominantly in Florida. It is considered by many to be among the finest tasting mangoes in the world; however, its poor yields have restrained the Edward from developing into a commercially significant variety.

The Edward was first propagated in the 1920s by Edward Simmonds of the Plant Introduction Garden in Miami, Florida and is believed to be a hybrid cross of Haden and Carabao mango cultivars.

Coenzyme A

Coenzyme A is phosphorylated to 4'-phosphopantothenate by the enzyme pantothenate kinase (PanK; CoaA; CoaX)
· A cysteine is added to 4'-phosphopantothenate by the enzyme phosphopantothenoylcysteine synthetase (PPC-DC; CoaB) to form 4'-phospho-N-pantothenoylcysteine (PPC)
· PPC is decarboxylated to 4'-phosphopantetheine by phosphopantothenoylcysteine decarboxylase (CoaC)
· 4'-phosphopantetheine is adenylylated to form dephospho-CoA by the enzyme phosphopantetheine adenylyl transferase (CoaD)
· Finally, dephospho-CoA is phosphorylated using ATP to Coenzyme A by the enzyme dephospho Coenzyme A kinase (CoaE.)

Since Coenzyme A is chemically a thiol, it can react with carboxylic acids to form thioesters, thus functioning as an acyl group carrier. It assists in transferring fatty acids from the cytoplasm to mitochondria.

Electron transport chain

An Electron transport chain couples a chemical reaction between an electron donor (such as NADH) and an electron acceptor (such as O_2) to the transfer of H^+ ions across a membrane, through a set of mediating biochemical reactions. These H^+ ions are used to produce adenosine triphosphate (ATP), the main energy intermediate in living organisms, as they move back across the membrane. Electron transport chain s are used for extracting energy from sunlight (photosynthesis) and from redox reactions such as the oxidation of sugars (respiration.)

Horse murders

The Horse murders scandal was a form of insurance fraud in the United States in which expensive horses, many of them show jumpers, were insured against death, accident and then killed to collect the insurance money. It is not known how many horses were murdered between the mid 1970s and the mid-1990s, when a Federal Bureau of Investigation (FBI) investigation brought the horse killings to light, but the number is thought to be well over 50, and may have been as high as 100. In addition, in 1977, the heiress Helen Brach disappeared and was presumed by law enforcement agents to have been murdered by the perpetrators of these crimes, because she threatened to report their criminal activity to authorities; continuing investigations into Brach's death began to uncover the insurance fraud in the 1990s.

Acid	An acid is traditionally considered any chemical compound that, when dissolved in water, gives a solution with a hydrogen ion activity greater than in pure water, i.e. a pH less than 7.0. That approximates the modern definition of Johannes Nicolaus Brønsted and Martin Lowry, who independently defined an acid as a compound which donates a hydrogen ion (H^+) to another compound (called a base.) Common examples include acetic acid and sulfuric acid (used in car batteries.)
Citric acid	Citric acid is a weak organic acid, and it is a natural preservative and is also used to add an acidic taste to foods and soft drinks. In biochemistry, it is important as an intermediate in the Citric acid cycle and therefore occurs in the metabolism of virtually all living things. It can also be used as an environmentally benign cleaning agent.
Citric acid cycle	The Citric acid cycle -- also known as the tricarboxylic acid cycle, the Krebs cycle the Szent-Györgyi-Krebs cycle -- is a series of enzyme-catalysed chemical reactions of central importance in all living cells that use oxygen as part of cellular respiration. In eukaryotes, the Citric acid cycle occurs in the matrix of the mitochondrion. The components and reactions of the Citric acid cycle were established by seminal work from Albert Szent-Györgyi and Hans Krebs.
Cristae	Cristae are the internal compartments formed by the inner membrane of a mitochondrion. They are studded with proteins, including ATP synthase and a variety of cytochromes. The maximum surface for chemical reactions to occur is within the mitochondria.
Intermembrane space	The Intermembrane space is the region between the inner membrane and the outer membrane of a mitochondrion or a chloroplast. The main function of the Intermembrane space is oxidative phosphorylation. Channel proteins called porins in the outer membrane allow free movement of ions and small molecules into the Intermembrane space.
Alagille syndrome	Alagille syndrome is a genetic disorder that affects the liver, heart, and other systems of the body. Problems associated with the disorder generally become evident in infancy or early childhood. The disorder is inherited in an autosomal dominant pattern, and the estimated prevalence of Alagille syndrome is 1 in every 100,000 live births.
Globular protein	Globular protein s comprising 'globe'-like proteins that are more or less soluble in aqueous solutions (where they form colloidal solutions.) This main characteristic helps distinguishing them from fibrous proteins (the other class), which are practically insoluble. The term globin can refer more specifically to proteins including the globin fold.
Metabolic pathway	In biochemistry, a Metabolic pathway is a series of chemical reactions occurring within a cell. In each pathway, a principal chemical is modified by chemical reactions. Enzymes catalyze these reactions, and often require dietary minerals, vitamins, and other cofactors in order to function properly.
Metabolism	Metabolism is the set of chemical reactions that occur in living organisms in order to maintain life. These processes allow organisms to grow and reproduce, maintain their structures, and respond to their environments. Metabolism is usually divided into two categories.

Y chromosome

The Y chromosome is the sex-determining chromosome in most mammals, including humans. In mammals, it contains the gene SRY, which triggers testis development, thus determining sex. The human Y chromosome is composed of about 60 million base pairs.

Fermentation

Fermentation is the process of deriving energy from the oxidation of organic compounds, such as carbohydrates, using an endogenous electron acceptor, which is usually an organic compound. This is in contrast to cellular respiration, where electrons are donated to an exogenous electron acceptor, such as oxygen, via an electron transport chain. Fermentation does not necessarily have to be carried out in an anaerobic environment.

Oxygen

Oxygen and -γενÎ®ς (-genÄ"s) (producer, literally begetter) is the element with atomic number 8 and represented by the symbol O. It is a member of the chalcogen group on the periodic table, and is a highly reactive nonmetallic period 2 element that readily forms compounds (notably oxides) with almost all other elements. At standard temperature and pressure two atoms of the element bind to form di Oxygen , a colorless, odorless, tasteless diatomic gas with the formula O_2. Oxygen is the third most abundant element in the universe by mass after hydrogen and helium and the most abundant element by mass in the Earth's crust.

Ascaris

Ascaris is a genus of parasitic nematode worms known as the giant intestinal roundworms. One species, A. suum, typically infects pigs, while another, A. lumbricoides, affects human populations, typically in sub-tropical and tropical areas with poor sanitation. A. lumbricoides is the largest intestinal roundworm and is the most common helminth infection of humans worldwide, an infection known as ascariasis.

Ascaris lumbricoides

Ascaris lumbricoides is the member of the Ascaris family responsible for the disease ascariasis.

It can reach a length of up to 35 cm.

Ascaris lumbricoides, or 'roundworm', infections in humans occur when an ingested infective egg releases a larval worm that penetrates the wall of the duodenum and enters the bloodstream.

Cell

The Cell is the structural and functional unit of all known living organisms. It is the smallest unit of an organism that is classified as living, and is often called the building block of life. Some organisms, such as most bacteria, are unicellular (consist of a single Cell.)

DNA

Deoxyribonucleic acid (DNA) is a nucleic acid that contains the genetic instructions used in the development and functioning of all known living organisms and some viruses. The main role of DNA molecules is the long-term storage of information. DNA is often compared to a set of blueprints or a recipe, or a code, since it contains the instructions needed to construct other components of cells, such as proteins and RNA molecules.

DNA replication

DNA replication, the basis for biological inheritance, is a fundamental process occurring in all living organisms to copy their DNA. This process is 'semiconservative' in that each strand of the original double-stranded DNA molecule serves as template for the reproduction of the complementary strand. Hence, following DNA replication, two identical DNA molecules have been produced from a single double-stranded DNA molecule. Cellular proofreading and error-checking mechanisms ensure near perfect fidelity for DNA replication.

Shigella

Shigella is a genus of Gram-negative, non-spore forming rod-shaped bacteria closely related to Escherichia coli and Salmonella. The causative agent of human shigellosis, Shigella cause disease in primates, but not in other mammals. It is only naturally found in humans and apes.

Shigella dysenteriae

Shigella dysenteriae is a species of the rod-shaped bacterial genus Shigella. Shigella can cause shigellosis (bacillary dysentery.) Shigellae are Gram-negative, non-spore-forming, facultatively anaerobic, non-motile bacteria.

Centromere

A Centromere is a region of DNA typically found near the middle of a chromosome where two identical sister chromatids come in contact. It is involved in cell division as the point of mitotic spindle.

The Centromere s are, together with telomeres and origins of replication, one of the essential parts of any eukaryotic chromosome.

Chromatid

A Chromatid is one among the two identical copies of DNA making up a replicated chromosome, which are joined at their centromeres, for the process of cell division (mitosis or meiosis.) The term is used so long as the centromeres remain in contact. When they separate (during anaphase of mitosis and anaphase 2 of meiosis), the strands are called daughter-chromosomes.

Chromatin	Chromatin is the complex combination of DNA, RNA, and protein that makes up chromosomes. It is found inside the nuclei of eukaryotic cells, and within the nucleoid in prokaryotic cells. It is divided between hetero Chromatin (condensed) and eu Chromatin (extended) forms.
Chromosome	A Chromosome is an organized structure of DNA and protein that is found in cells. It is a single piece of coiled DNA containing many genes, regulatory elements and other nucleotide sequences. Chromosome s also contain DNA-bound proteins, which serve to package the DNA and control its functions.
Histones	In biology, Histones are the chief protein components of chromatin. They act as spools around which DNA winds, and they play a role in gene regulation. Without Histones, the unwound DNA in chromosomes would be very long.
Cell cycle	The Cell cycle is the series of events that take place in a cell leading to its division and duplication (replication.) In cells without a nucleus (prokaryotes), the Cell cycle occurs via a process termed binary fission. In cells with a nucleus (eukaryotes), the Cell cycle can be divided in two brief periods: interphase--during which the cell grows, accumulating nutrients needed for mitosis and duplicating its DNA--and the mitosis phase, during which the cell splits itself into two distinct cells, often called 'daughter cells'.
Cystic fibrosis	Cystic fibrosis is a genetic disorder known to be an inherited disease of the secretory glands, including the glands that make mucus and sweat. The hallmarks of Cystic fibrosis are salty tasting skin, normal appetite but poor growth and poor weight gain, excess mucus production, and coughing/shortness of breath. Males can be infertile due to the condition congenital bilateral absence of the vas deferens.
Cytokinesis	Cytokinesis is the process in which the cytoplasm of a single eukaryotic cell is divided to form two daughter cells. It usually initiates during the late stages of mitosis, and sometimes meiosis, splitting a binucleate cell in two, to ensure that chromosome number is maintained from one generation to the next. In animal cells, one notable exception to the normal process of Cytokinesis is oogenesis (the creation of an ovum in the ovarian follicle of the ovary), where the ovum takes almost all the cytoplasm and organelles, leaving very little for the resulting polar bodies, which then die.
Interphase	Interphase is the phase of the cell cycle in which the cell spends the majority of its time and performs the majority of its purposes including preparation for cell division. In preparation for cell division it increases its size and number of organelles, and makes a copy of its DNA. Interphase is also considered to be the 'living' phase of the cell, in which the cell obtains nutrients, grows, reads its DNA, and conducts other 'normal' cell functions. The majority of eukaryotic cells spend most of their time in Interphase.
Mitosis	Mitosis is the process in which a eukaryotic cell separates the chromosomes in its cell nucleus into two identical sets in two daughter nuclei. It is generally followed immediately by cytokinesis, which divides the nuclei, cytoplasm, organelles and cell membrane into two daughter cells containing roughly equal shares of these cellular components. mitosis and cytokinesis together define the mitotic (M) phase of the cell cycle - the division of the mother cell into two daughter cells, genetically identical to each other and to their parent cell.

Anaphase

Anaphase, is from the ancient Greek á¼€vî¬ and φî¬σις (stage), is the stage of mitosis when chromosomes separate in a eukaryotic cell. Each chromatid moves to opposite poles of the cell, the opposite ends of the mitotic spindle, near the microtubule organizing centers. During this stage, Anaphase lag could happen.

Aster

An Aster is a cellular structure shaped like a star, formed around each centrosome during mitosis in an animal cell. Astral rays, composed of microtubules, radiate from the centrosphere and appear as a cloud.

Centrosome

In cell biology, the Centrosome is an organelle that serves as the main microtubule organizing center (MTOC) of the animal cell as well as a regulator of cell-cycle progression. It was discovered by Edouard Van Beneden in 1883 and was described and named in 1888 by Theodor Boveri. The Centrosome is thought to have evolved only in the metazoan lineage of eukaryotic cells.

Metaphase

Metaphase from the ancient Greek μετî¬ and φî¬σις (stage), is a stage of mitosis in the eukaryotic cell cycle in which condensed ' highly coiled chromosomes, carrying genetic information, align in the middle of the cell before being separated into each of the two daughter cells.

Preceded by events in pro Metaphase and followed by anaphase, microtubules formed in prophase have already found and attached themselves to kinetochores in Metaphase The centromeres of the chromosomes convene themselves on the Metaphase plate (or equatorial plate), an imaginary line that is equidistant from the two centrosome poles.

Prophase

Prophase is a stage of mitosis in which the chromatin condenses into a highly ordered structure called a chromosome (it is at this stage giemsa staining can be applied to elicit G-banding in chromosomes) in which the chromatin becomes visible. This process, called chromatin condensation, is mediated by the condensin complex. Since the genetic material has been duplicated in an earlier phase of the cell cycle, there are two identical copies of each chromosome in the cell.

Cell plate

Cytokinesis in terrestrial plants occurs by Cell plate formation. This process entails the delivery of Golgi-derived and endosomal vesicles carrying cell wall and cell membrane components to the plane of cell division and the subsequent fusion of these vesicles within this plane.

After formation of an early tubulo-vesicular network at the center of the cell, the initially labile Cell plate consolidates into a tubular network and eventually a fenestrated sheet.

CVS

CVS is a terpene cyclase enzyme responsible for the biosynthesis of valencene, a sesquiterpene, using farnesyl pyrophosphate as its substrate. The first CVS enzyme was isolated using orange cDNA. .

Cycads

Cycads are a group of seed plants characterized by a large crown of compound leaves and a stout trunk. They are evergreen, gymnospermous, dioecious plants having large pinnately compound leaves. They are frequently confused with and mistaken for palms or ferns, but are related to neither, belonging to the division Cycadophyta.

Cyclin

Cyclin s are a family of proteins which control the progression of cells through the cell cycle by activating Cyclin dependent kinase (Cdk) enzymes. Expression of human Cyclin s through the cell cycle.

Cyclin s are so named because their concentration varies in a cyclical fashion during the cell cycle; they are produced or degraded as needed in order to drive the cell through the different stages of the cell cycle.

A Cyclin forms a complex with Cdk.

Kinase

In chemistry and biochemistry, a kinase alternatively known as a phosphotransferase, is a type of enzyme that transfers phosphate groups from high-energy donor molecules, such as ATP, to specific substrates. The process is referred to as phosphorylation. An enzyme that removes phosphate groups is known as a phosphatase.

APC

APC (adenomatosis polyposis coli) is a human gene that is classified as a tumor suppressor gene. Tumor suppressor genes prevent the uncontrolled growth of cells that may result in cancerous tumors. The protein made by the APC gene plays a critical role in several cellular processes that determine whether a cell may develop into a tumor.

Apoptosis

Apoptosis is the process of programmed cell death that may occur in multicellular organisms. Programmed cell death involves a series of biochemical events leading to a characteristic cell morphology and death, in more specific terms, a series of biochemical events that lead to a variety of morphological changes, including blebbing, changes to the cell membrane such as loss of membrane asymmetry and attachment, cell shrinkage, nuclear fragmentation, chromatin condensation, and chromosomal DNA fragmentation (1-4.)

Contact inhibition

Contact inhibition is the natural process of arresting cell growth when two or more cells come into contact with each other. Oncologists use this property to distinguish between normal and cancerous cells.

Cell lines used widely in animal cell culture laboratories are genetically modified to suppress apoptosis however maintain this phenomenon.

Telomere

A Telomere is a region of repetitive DNA at the end of chromosomes, which protects the end of the chromosome from destruction.

A Russian theorist Alexei Olovnikov was the first to recognize the problem of how chromosomes could replicate right to the tip, as such was impossible with replication in a 3' to 5' direction. To solve this and to accommodate Leonard Hayflick's idea of limited somatic cell division, Olovnikov suggested that DNA sequences would be lost in every replicative phase until they reached a critical level, at which point cell division would stop.

Crassulacean acid metabolism

Crassulacean acid metabolism is an elaborate carbon fixation pathway in some plants. These plants fix carbon dioxide during the night, storing it as the four carbon acid malate. The CO_2 is released during the day, where it is concentrated around the enzyme RuBisCO, increasing the efficiency of photosynthesis.

Cancer

Cancer is a genetic disorder in which the normal control of cell growth is lost. Cancer genetics is now one of the fastest expanding medical specialties. At the molecular level, Cancer is caused by mutation(s) in DNA, which result in aberrant cell proliferation.

Tumor

A tumor or tumour is the name for a swelling or lesion formed by an abnormal growth of cells (termed neoplastic.) tumor is not synonymous with cancer. A tumor can be benign, pre-malignant or malignant, whereas cancer is by definition malignant.

Punnett square

The Punnett square is a diagram that is used to predict the outcome of a particular cross or breeding experiment. It is named after Reginald C. Punnett, who devised the approach, and is used by biologists to determine the probability of an offspring having a particular genotype. The Punnett square is a summary of every possible combination of one maternal allele with one paternal allele for each gene being studied in the cross.

Sulfolobus

In taxonomy, Sulfolobus is a genus of the Sulfolobaceae.

	Sulfolobus species grow in volcanic springs with optimal growth occurring at pH 2-3 and temperatures of 75-80 °C, making them acidophiles and thermophiles respectively. Sulfolobus cells are irregularly shaped and flagellar.
Radiation	In physics, radiation describes any process in which energy emitted by one body travels through a medium or through space, ultimately to be absorbed by another body. Non-physicists often associate the word with ionizing radiation, but it can also refer to electromagnetic radiation (i.e., radio waves, infrared light, visible light, ultraviolet light, and X-rays) which can also be ionizing radiation, to acoustic radiation, or to other more obscure processes. What makes it radiation is that the energy radiates (i.e., it travels outward in straight lines in all directions) from the source.
Sugar	Sugar is a class of edible crystalline substances, mainly sucrose, lactose, and fructose. Human taste buds interpret its flavor as sweet. Sugar as a basic food carbohydrate primarily comes from Sugar cane and from Sugar beet, but also appears in fruit, honey, sorghum, Sugar maple (in maple syrup), and in many other sources.

DNA

Deoxyribonucleic acid (DNA) is a nucleic acid that contains the genetic instructions used in the development and functioning of all known living organisms and some viruses. The main role of DNA molecules is the long-term storage of information. DNA is often compared to a set of blueprints or a recipe, or a code, since it contains the instructions needed to construct other components of cells, such as proteins and RNA molecules.

Cell

The Cell is the structural and functional unit of all known living organisms. It is the smallest unit of an organism that is classified as living, and is often called the building block of life. Some organisms, such as most bacteria, are unicellular (consist of a single Cell.)

Meiosis

In biology, Meiosis is a process of reductional division in which the number of chromosomes per cell is halved. In animals, Meiosis always results in the formation of gametes, while in other organisms it can give rise to spores. As with mitosis, before Meiosis begins, the DNA in the original cell is replicated during S-phase of the cell cycle.

Technology

Technology is a broad concept that deals with an animal species' ethology or behavior of usage and of knowledge of tools and crafts, and how it affects the animal species' ability to control and adapt to its environment. Technology is a term with origins in the Greek 'technologia', 'τεχνολογΐα' -- 'techne', 'τῖχνη' and 'logia', 'λογΐα' ('saying'.) However, a strict definition is elusive; 'Technology' can refer to material objects of use to humanity, such as machines, hardware or utensils, but can also encompass broader themes, including systems, methods of organization, and techniques.

X chromosome

The X chromosome is one of the two sex-determining chromosomes in many animal species, including mammals (the other is the Y chromosome.) It is a part of the XY sex-determination system and X0 sex-determination system. The X chromosome was named for its unique properties by early researchers, and this resulted in its counterpart being named the Y chromosome for the next letter in the alphabet when it was discovered later.

Y chromosome

The Y chromosome is the sex-determining chromosome in most mammals, including humans. In mammals, it contains the gene SRY, which triggers testis development, thus determining sex. The human Y chromosome is composed of about 60 million base pairs.

Alagille syndrome

Alagille syndrome is a genetic disorder that affects the liver, heart, and other systems of the body. Problems associated with the disorder generally become evident in infancy or early childhood. The disorder is inherited in an autosomal dominant pattern, and the estimated prevalence of Alagille syndrome is 1 in every 100,000 live births.

Genome

In classical genetics, the Genome of a diploid organism including eukarya refers to a full set of chromosomes or genes in a gamete; thereby, a regular somatic cell contains two full sets of Genome s. In haploid organisms, including bacteria, archaea, viruses, and mitochondria, a cell contains only a single set of the Genome usually in a single circular or contiguous linear DNA (or RNA for retroviruses.) In modern molecular biology the Genome of an organism is its hereditary information encoded in DNA (or, for retroviruses, RNA.)

Genome Project

Genome project s are scientific endeavours that ultimately aim to determine the complete genome sequence of an organism (be it an animal, a plant, a fungus, a bacterium, an archaean, a protist or a virus.) The genome sequence for any organism requires the DNA sequences for each of the chromosomes in an organism to be determined. For bacteria, which usually have just one chromosome, a Genome project will aim to map the sequence of that chromosome.

Human
A Human is a member of a species of bipedal primates in the family Hominidae . DNA and fossil evidence indicates that modern Human s originated in east Africa about 200,000 years ago. When compared to other animals and primates, Human s have a highly developed brain, capable of abstract reasoning, language, introspection and problem solving.

Human Genome
The Human genome is the genome of Homo sapiens, which is stored on 23 chromosome pairs. Twenty-two of these are autosomal chromosome pairs, while the remaining pair is sex-determining. The haploid Human genome occupies a total of just over 3 billion DNA base pairs.

Human Genome Project
The Human Genome Project was an international scientific research project with a primary goal to determine the sequence of chemical base pairs which make up DNA and to identify and map the approximately 20,000-25,000 genes of the human genome from both a physical and functional standpoint.

The project began in 1990 initially headed by James D. Watson at the U.S. National Institutes of Health. A working draft of the genome was released in 2000 and a complete one in 2003, with further analysis still being published.

Allele
An Allele) is one of a series of different forms of a gene. The word is a short from of allelomorph , which was used in the early days of genetics to describe variant forms of a gene detected as different phenotypes. Allele s are now understood to be alternative DNA sequences at the same physical gene locus, which may or may not result in different phenotypic traits.

Horse murders
The Horse murders scandal was a form of insurance fraud in the United States in which expensive horses, many of them show jumpers, were insured against death, accident and then killed to collect the insurance money. It is not known how many horses were murdered between the mid 1970s and the mid-1990s, when a Federal Bureau of Investigation (FBI) investigation brought the horse killings to light, but the number is thought to be well over 50, and may have been as high as 100. In addition, in 1977, the heiress Helen Brach disappeared and was presumed by law enforcement agents to have been murdered by the perpetrators of these crimes, because she threatened to report their criminal activity to authorities; continuing investigations into Brach's death began to uncover the insurance fraud in the 1990s.

Leukemia inhibitory factor
Leukemia inhibitory factor an interleukin 6 class cytokine, is a chemical in cells that affects their growth and development.

Leukemia inhibitory factor derives its name from its ability to induce the terminal differentiation of myeloid leukaemic cells. Other properties attributed to the cytokine include: the growth promotion and cell differentiation of different types of target cells, influence on bone metabolism, cachexia, neural development, embryogenesis and inflammation.

Oogenesis
Oogenesis or oögenesis is the creation of an ovum . It is the female process of gametogenesis. It involves the various stages of immature ova.

Spermatogenesis
Spermatogenesis is the process by which male spermatogonia develop into mature spermatozoa. Spermatozoa are the mature male gametes in many sexually reproducing organisms. Thus, Spermatogenesis is the male version of gametogenesis.

Zygote

A zygote is a term in Developmental biology used to describe the first stage of a new unique organism when it consists of just a single cell. The term is also used more loosely to refer to the group of cells formed by the first few cell divisions, although this is properly referred to as a blastomere. A zygote is usually produced by a fertilization event between two haploid cells - an ovum from a female and a sperm cell from a male - which combine to form the single diploid cell.

Dyad

A Dyad is a pair of sister chromatids. This occurs in prophase I of meiosis. After DNA replication, the two sister chromatids align side-by-side and appear to have an undivided centromere, in contrast to mitosis, in which each chromatid appears to have its own separate centromere.

Synapsis

Synapsis is the pairing of two homologous chromosomes that occurs during meiosis. It is a form of chromosomal crossover. Synapsis takes place during prophase I. When homologous chromosomes synapse, they come closer together until they are connected by a protein complex called the synaptonemal complex, which contains central and lateral elements.

Chromatid

A Chromatid is one among the two identical copies of DNA making up a replicated chromosome, which are joined at their centromeres, for the process of cell division (mitosis or meiosis.) The term is used so long as the centromeres remain in contact. When they separate (during anaphase of mitosis and anaphase 2 of meiosis), the strands are called daughter-chromosomes.

Mitosis

Mitosis is the process in which a eukaryotic cell separates the chromosomes in its cell nucleus into two identical sets in two daughter nuclei. It is generally followed immediately by cytokinesis, which divides the nuclei, cytoplasm, organelles and cell membrane into two daughter cells containing roughly equal shares of these cellular components. mitosis and cytokinesis together define the mitotic (M) phase of the cell cycle - the division of the mother cell into two daughter cells, genetically identical to each other and to their parent cell.

Trisomy

A Trisomy is a genetic abnormality in which there are three copies, instead of the normal two, of a particular chromosome. Most organisms that reproduce sexually have pairs of chromosomes in each cell, with one chromosome inherited from each parent. In such organisms, a process called meiosis creates cells called gametes (eggs or sperm) that have only one set of chromosomes.

Barr body

In those species (including humans) in which sex is determined by the presence of the Y or W chromosome rather than the diploidy of the X or Z, a Barr body is the inactive X chromosome in a female cell 2003), rendered inactive in a process called Lyonization. The Lyon hypothesis states that in cells with multiple X chromosomes, all but one are inactivated during mammalian embryogenesis (Lyon, 1961.) This happens early in embryonic development at random in mammals, (Brown, 1997) except in marsupials and in some extra-embryonic tissues of some placental mammals, in which the father's X chromosome is always deactivated (Lee, 2003.)

Shigella	Shigella is a genus of Gram-negative, non-spore forming rod-shaped bacteria closely related to Escherichia coli and Salmonella. The causative agent of human shigellosis, Shigella cause disease in primates, but not in other mammals. It is only naturally found in humans and apes.
Shigella dysenteriae	Shigella dysenteriae is a species of the rod-shaped bacterial genus Shigella. Shigella can cause shigellosis (bacillary dysentery.) Shigellae are Gram-negative, non-spore-forming, facultatively anaerobic, non-motile bacteria.
Cell	The Cell is the structural and functional unit of all known living organisms. It is the smallest unit of an organism that is classified as living, and is often called the building block of life. Some organisms, such as most bacteria, are unicellular (consist of a single Cell.)
Cystic fibrosis	Cystic fibrosis is a genetic disorder known to be an inherited disease of the secretory glands, including the glands that make mucus and sweat. The hallmarks of Cystic fibrosis are salty tasting skin, normal appetite but poor growth and poor weight gain, excess mucus production, and coughing/shortness of breath. Males can be infertile due to the condition congenital bilateral absence of the vas deferens.
Horse murders	The Horse murders scandal was a form of insurance fraud in the United States in which expensive horses, many of them show jumpers, were insured against death, accident and then killed to collect the insurance money. It is not known how many horses were murdered between the mid 1970s and the mid-1990s, when a Federal Bureau of Investigation (FBI) investigation brought the horse killings to light, but the number is thought to be well over 50, and may have been as high as 100. In addition, in 1977, the heiress Helen Brach disappeared and was presumed by law enforcement agents to have been murdered by the perpetrators of these crimes, because she threatened to report their criminal activity to authorities; continuing investigations into Brach's death began to uncover the insurance fraud in the 1990s.
Genetic	Genetics is the study of how living things receive common traits from previous generations. These traits are described by the Genetic information carried by a molecule called DNA. The instructions for constructing and operating an organism are contained in the organism's DNA. Every living thing on earth has DNA in its cells. Genes are the hereditary components of DNA that occupy spots on chromosomes and determine characteristics in an organism.
Cancer	Cancer is a genetic disorder in which the normal control of cell growth is lost. Cancer genetics is now one of the fastest expanding medical specialties. At the molecular level, Cancer is caused by mutation(s) in DNA, which result in aberrant cell proliferation.
Nutrient	A Nutrient is a chemical that an organism needs to live and grow or a substance used in an organism's metabolism which must be taken in from its environment. Nutrient s are the substances that enrich the body. They build and repair tissues, give heat and energy, and regulate body processes.
Punnett square	The Punnett square is a diagram that is used to predict the outcome of a particular cross or breeding experiment. It is named after Reginald C. Punnett, who devised the approach, and is used by biologists to determine the probability of an offspring having a particular genotype. The Punnett square is a summary of every possible combination of one maternal allele with one paternal allele for each gene being studied in the cross.

DNA

Deoxyribonucleic acid (DNA) is a nucleic acid that contains the genetic instructions used in the development and functioning of all known living organisms and some viruses. The main role of DNA molecules is the long-term storage of information. DNA is often compared to a set of blueprints or a recipe, or a code, since it contains the instructions needed to construct other components of cells, such as proteins and RNA molecules.

Allele

An Allele) is one of a series of different forms of a gene. The word is a short from of allelomorph , which was used in the early days of genetics to describe variant forms of a gene detected as different phenotypes. Allele s are now understood to be alternative DNA sequences at the same physical gene locus, which may or may not result in different phenotypic traits.

Gene

A Gene is the basic unit of heredity in a living organism. All living things depend on Gene s. Gene s hold the information to build and maintain their cells and pass Gene tic traits to offspring.

Heterochromatin

Heterochromatin is a tightly packed form of DNA. Its major characteristic is that transcription is limited. As such, it is a means to control gene expression, through regulation of the transcription initiation.

Chromatin is found in two varieties: euchromatin and Heterochromatin.

Heterozygous

Zygosity refers to the similarity of genes for a trait (inherited characteristic) in an organism. If both genes are the same, the organism is homozygous for the trait. If both genes are different, the organism is heterozygous for that trait.

Locus

In the fields of genetics and evolutionary computation, a Locus is a fixed position on a chromosome such as the position of a biomarker that may be occupied by one or more genes. A variant of the DNA sequence at a given Locus is called an allele. The ordered list of loci known for a particular genome is called a genetic map.

Phenotype

A Phenotype is any observable characteristic or trait of an organism: such as its morphology, development, biochemical or physiological properties, or behavior. Phenotype s result from the expression of an organism's genes as well as the influence of environmental factors and possible interactions between the two. The genotype of an organism is the inherited instructions it carries within its genetic code.

Technology

Technology is a broad concept that deals with an animal species' ethology or behavior of usage and of knowledge of tools and crafts, and how it affects the animal species' ability to control and adapt to its environment. Technology is a term with origins in the Greek 'technologia', 'τεχνολογΙ¯α' -- 'techne', 'τΙ̂χνη' and 'logia', 'λογΙ¯α' ('saying'.) However, a strict definition is elusive; 'Technology' can refer to material objects of use to humanity, such as machines, hardware or utensils, but can also encompass broader themes, including systems, methods of organization, and techniques.

Drosophila

Drosophila has long been a favorite model system for geneticists and developmental biologists studying embryogenesis. The small size, short generation time, and large brood size makes it ideal for genetic studies. Transparent embryos facilitate developmental studies.

Fly

Nematocera (includes Eudiptera)Brachycera

True flies are insects of the order Diptera , possessing a single pair of wings on the mesothorax and a pair of halteres, derived from the hind wings, on the metathorax.

The presence of a single pair of wings distinguishes true flies from other insects with 'Fly' in their name, such as mayflies, dragonflies, damselflies, stoneflies, whiteflies, fireflies, alderflies, dobsonflies, snakeflies, sawflies, caddisflies, butterflies or scorpionflies. Some true flies have become secondarily wingless, especially in the superfamily Hippoboscoidea, or among those that are inquilines in social insect colonies.

Fruit

The term fruit has different meanings dependent on context, and the term is not synonymous in food preparation and biology. fruit s are the means by which flowering plants disseminate seeds, and the presence of seeds indicates that a structure is most likely a fruit though not all seeds come from fruit s.

No single terminology really fits the enormous variety that is found among plant fruit s.

Virus

A virus is a microscopic infectious agent that can reproduce only inside a host cell. virus es infect all types of organisms: from animals and plants, to bacteria and archaea. Since the initial discovery of tobacco mosaic virus by Martinus Beijerinck in 1898, more than 5,000 types of virus have been described in detail, although most types of virus remain undiscovered.

Maintenance

Maintenance of an organism is the collection of processes to stay alive, excluding production processes. The Dynamic Energy Budget theory delineates two classes

· Somatic Maintenance. This comprises the turnover of structural mass (mainly proteins), the Maintenance of concentration gradients of metabolites across membranes, activity
· Maturity Maintenance. This comprises the Maintenance of defence systems (such as the immune system), the preparation of the body for reproduction.
The theory assumes that maturity Maintenance costs can be reduced more easily during starvation than somatic Maintenance costs. Under extreme starvation conditions, somatic Maintenance costs are paid from structural mass, which causes shrinking.Some organism manage to switch to the turpor state under starvation conditions, and reduce their Maintenance costs.

Meiosis

In biology, Meiosis is a process of reductional division in which the number of chromosomes per cell is halved. In animals, Meiosis always results in the formation of gametes, while in other organisms it can give rise to spores. As with mitosis, before Meiosis begins, the DNA in the original cell is replicated during S-phase of the cell cycle.

Blood type

A Blood type is a classification of blood based on the presence or absence of inherited antigenic substances on the surface of red blood cells These antigens may be proteins, carbohydrates, glycoproteins depending on the blood group system, and some of these antigens are also present on the surface of other types of cells of various tissues. Several of these red blood cell surface antigens, that stem from one allele, collectively form a blood group system.

Dominance

In genetics, Dominance describes the effects of the different versions of a particular gene on the phenotype of an organism. Many animals (including humans) and plants have two copies of each gene in their genome, one inherited from each parent. The different variants of a specific gene (such as that coding for earlobes) are known as alleles.

Trait

A trait is a distinct variant of a phenotypic character of an organism that may be inherited, environmentally determined or somewhere in between. For example, eye color is a character or abstraction of an attribute, while blue, brown and hazel are traits.

	A trait may be any single feature or quantifiable measurement of an organism.
X chromosome	The X chromosome is one of the two sex-determining chromosomes in many animal species, including mammals (the other is the Y chromosome.) It is a part of the XY sex-determination system and X0 sex-determination system. The X chromosome was named for its unique properties by early researchers, and this resulted in its counterpart being named the Y chromosome for the next letter in the alphabet when it was discovered later.
Y chromosome	The Y chromosome is the sex-determining chromosome in most mammals, including humans. In mammals, it contains the gene SRY, which triggers testis development, thus determining sex. The human Y chromosome is composed of about 60 million base pairs.
Human	A Human is a member of a species of bipedal primates in the family Hominidae . DNA and fossil evidence indicates that modern Human s originated in east Africa about 200,000 years ago. When compared to other animals and primates, Human s have a highly developed brain, capable of abstract reasoning, language, introspection and problem solving.
Gamete	A Gamete is a cell that fuses with another Gamete during fertilization (conception) in organisms that reproduce sexually. In species that produce two morphologically distinct types of Gamete s, and in which each individual produces only one type, a female is any individual that produces the larger type of Gamete -- called an ovum (or egg) -- and a male produces the smaller tadpole-like type -- called a sperm. This is an example of anisogamy or heterogamy, the condition wherein females and males produce Gamete s of different sizes (this is the case in humans; the human ovum is approximately 20 times larger than the human sperm cell.)

Decomposers

Decomposers are organisms that consume dead or decaying organisms, and, in doing so, carry out the natural process of decomposition. Like herbivores and predators, Decomposers are heterotrophic, meaning that they use organic substrates to get their energy, carbon and nutrients for growth and development. Decomposers use deceased organisms and non-living organic compounds as their food source.

Virginia opossum

The Virginia Opossum, commonly known as the North American Opossum, is the only marsupial found in North America north of the Rio Grande River. A solitary and nocturnal animal about the size of a domestic cat, it is a successful opportunist and is found throughout Central America and North America east of the Rockies from Costa Rica to southern Ontario (it was also introduced to California in 1910, and now occupies much of the Pacific coast); it seems to be still expanding its range northward. Its ancestors evolved in South America, but were enabled to invade North America in the Great American Interchange by the formation of the Isthmus of Panama about 3 million years ago.

Virus

A virus is a microscopic infectious agent that can reproduce only inside a host cell. virus es infect all types of organisms: from animals and plants, to bacteria and archaea. Since the initial discovery of tobacco mosaic virus by Martinus Beijerinck in 1898, more than 5,000 types of virus have been described in detail, although most types of virus remain undiscovered.

Cystic fibrosis

Cystic fibrosis is a genetic disorder known to be an inherited disease of the secretory glands, including the glands that make mucus and sweat.

The hallmarks of Cystic fibrosis are salty tasting skin, normal appetite but poor growth and poor weight gain, excess mucus production, and coughing/shortness of breath. Males can be infertile due to the condition congenital bilateral absence of the vas deferens.

Horse murders

The Horse murders scandal was a form of insurance fraud in the United States in which expensive horses, many of them show jumpers, were insured against death, accident and then killed to collect the insurance money. It is not known how many horses were murdered between the mid 1970s and the mid-1990s, when a Federal Bureau of Investigation (FBI) investigation brought the horse killings to light, but the number is thought to be well over 50, and may have been as high as 100. In addition, in 1977, the heiress Helen Brach disappeared and was presumed by law enforcement agents to have been murdered by the perpetrators of these crimes, because she threatened to report their criminal activity to authorities; continuing investigations into Brach's death began to uncover the insurance fraud in the 1990s.

Gene

A Gene is the basic unit of heredity in a living organism. All living things depend on Gene s. Gene s hold the information to build and maintain their cells and pass Gene tic traits to offspring.

Nucleotide

Nucleotide s are molecules that, when joined together, make up the structural units of RNA and DNA. Additionally, Nucleotide s play central roles in metabolism. In that capacity, they serve as sources of chemical energy (adenosine triphosphate and guanosine triphosphate), participate in cellular signaling (cyclic guanosine monophosphate and cyclic adenosine monophosphate), and are incorporated into important cofactors of enzymatic reactions (coenzyme A, flavin adenine di Nucleotide , flavin mono Nucleotide , and nicotinamide adenine di Nucleotide phosphate.) Figure 1: Structural elements of the most common Nucleotide s Figure 2: Ribose structure indicating numbering of carbon atoms

A Nucleotide is composed of a nucleobase (nitrogenous base) and a five-carbon sugar (either ribose or 2'-deoxyribose), and one to three phosphate groups.

DNA	Deoxyribonucleic acid (DNA) is a nucleic acid that contains the genetic instructions used in the development and functioning of all known living organisms and some viruses. The main role of DNA molecules is the long-term storage of information. DNA is often compared to a set of blueprints or a recipe, or a code, since it contains the instructions needed to construct other components of cells, such as proteins and RNA molecules.
Technology	Technology is a broad concept that deals with an animal species' ethology or behavior of usage and of knowledge of tools and crafts, and how it affects the animal species' ability to control and adapt to its environment. Technology is a term with origins in the Greek 'technologia', 'τεχνολογΓα' -- 'techne', 'τΐχνη' and 'logia', 'λογΓα' ('saying'.) However, a strict definition is elusive; 'Technology' can refer to material objects of use to humanity, such as machines, hardware or utensils, but can also encompass broader themes, including systems, methods of organization, and techniques.
DNA ligase	In molecular biology, DNA ligase is a special type of ligase (EC 6.5.1.1) that can link together two DNA strands that have double-strand break (a break in both complementary strands of DNA.) The alternative, a single-strand break, is fixed by a different type of DNA ligase using the complementary strand as a template but still requires DNA ligase to create the final phosphodiester bond to fully repair the DNA. DNA ligase has applications in both DNA repair and DNA replication In addition, DNA ligase has extensive use in molecular biology laboratories for Genetic recombination experiments
DNA polymerase	A DNA polymerase is an enzyme that catalyzes the polymerization of deoxyribonucleotides into a DNA strand. DNA polymerase s are best-known for their role in DNA replication, in which the polymerase 'reads' an intact DNA strand as a template and uses it to synthesize the new strand. The newly-polymerized molecule is complementary to the template strand and identical to the template's original partner strand.
DNA replication	DNA replication, the basis for biological inheritance, is a fundamental process occurring in all living organisms to copy their DNA. This process is 'semiconservative' in that each strand of the original double-stranded DNA molecule serves as template for the reproduction of the complementary strand. Hence, following DNA replication, two identical DNA molecules have been produced from a single double-stranded DNA molecule. Cellular proofreading and error-checking mechanisms ensure near perfect fidelity for DNA replication.
Escherichia	Escherichia is a genus of Gram-negative, non-spore forming, facultatively anaerobic, rod-shaped bacteria from the family Enterobacteriaceae. Inhabitants of the gastrointestinal tracts of warm-blooded animals, Escherichia species provide a portion of the microbially-derived vitamin K for their host. While many Escherichia are harmless commensals, particular strains of some species are human pathogens, and are known as the most common cause of urinary tract infections, significant sources of gastrointestinal disease, ranging from simple diarrhea to dysentery-like conditions, as well as a wide-range of other pathogenic states.
Escherichia coli	Escherichia coli , is a Gram negative bacterium that is commonly found in the lower intestine of warm-blooded organisms . Most E. coli strains are harmless, but some, such as serotype O157:H7, can cause serious food poisoning in humans, and are occasionally responsible for costly product recalls. The harmless strains are part of the normal flora of the gut, and can benefit their hosts by producing vitamin K_2, or by preventing the establishment of pathogenic bacteria within the intestine.
Semiconservative replication	Semiconservative replication describes the method by which DNA is replicated in all known cells. This method of replication was one of three proposed models of DNA replication:

· Conservative replication would leave the two original template DNA strands together in a double helix and would produce a copy composed of two new strands containing all of the new DNA base pairs.
· Dispersive replication would produce two copies of the DNA, both containing distinct regions of DNA composed of either both original strands or both new strands.
· Semiconservative replication would produce two copies that each contained one of the original strands and one entirely new strand.

The deciphering of the structure of DNA by Watson and Crick in 1953 suggested that each strand of the double helix would serve as a template for synthesis of a new strand. However, there was no way of guessing how the newly synthesized strands might combine with the template strands to form two double helical DNA molecules. The semiconservative model seemed most reasonable since it would allow each daughter strand to remain associated with its template strand.

RNA

Ribonucleic acid (RNA) is a biologically important type of molecule that consists of a long chain of nucleotide units. Each nucleotide consists of a nitrogenous base, a ribose sugar, and a phosphate. RNA is very similar to DNA, but differs in a few important structural details: in the cell, RNA is usually single-stranded, while DNA is usually double-stranded; RNA nucleotides contain ribose while DNA contains deoxyribose (a type of ribose that lacks one oxygen atom); and RNA has the base uracil rather than thymine that is present in DNA.

RNA is transcribed from DNA by enzymes called RNA polymerases and is generally further processed by other enzymes.

Messenger ribonucleic acid

Messenger ribonucleic acid is a molecule of RNA encoding a chemical 'blueprint' for a protein product. mRNA is transcribed from a DNA template, and carries coding information to the sites of protein synthesis: the ribosomes. Here, the nucleic acid polymer is translated into a polymer of amino acids: a protein.

Ribosomal RNA

Ribosomal RNA is the central component of the ribosome, the protein manufacturing machinery of all living cells. The function of the rRNA is to provide a mechanism for decoding mRNA into amino acids and to interact with the tRNAs during translation by providing peptidyl transferase activity.The tRNA then brings the necessary amino acids corresponding to the appropriate mRNA codon.

The ribosome is composed of two subunits P, and E.

· The A site in the ribosome binds to an aminoacyl-tRNA (a tRNA bound to an amino acid.)
· The amino (NH2) group of the aminoacyl-tRNA, which contains the new amino acid, attacks the ester linkage of peptidyl-tRNA (contained within the P site), which contains the last amino acid of the growing chain, forming a new peptide bond. This reaction is catalyzed by peptidyl transferase.
· The tRNA that was holding on the last amino acid is moved to the E site, and what used to be the aminoacyl-tRNA is now the peptidyl-tRNA.

A single mRNA can be translated simultaneously by multiple ribosomes.

Both prokaryotic and eukaryotic can be broken down into two subunits (the S in 16S represents Svedberg units):

Note that the S units of the subunits cannot simply be added because they represent measures of sedimentation rate rather than of mass.

Ribosome	Ribosome s are complexes of RNA and protein that are found in all cells with nuclei. Ribosome s from bacteria, archaea and eukaryotes (the three domains of life on Earth), have significantly different structure and RNA. The Ribosome s in the mitochondria of eukaryotic cells resemble those in bacteria, reflecting the evolutionary origin of this organelle. The Ribosome functions in the expression of the genetic code from nucleic acid into protein, in a process called translation.
Transfer RNA	Transfer RNA is a small RNA molecule (usually about 74-95 nucleotides) that transfers a specific active amino acid to a growing polypeptide chain at the ribosomal site of protein synthesis during translation. It has a 3' terminal site for amino acid attachment. This covalent linkage is catalyzed by an aminoacyl tRNA synthetase.
Edward	The Edward mango is a monoembryonic mango cultivar grown predominantly in Florida. It is considered by many to be among the finest tasting mangoes in the world; however, its poor yields have restrained the Edward from developing into a commercially significant variety. The Edward was first propagated in the 1920s by Edward Simmonds of the Plant Introduction Garden in Miami, Florida and is believed to be a hybrid cross of Haden and Carabao mango cultivars.
Gene expression	Gene expression is the process by which information from a gene is used in the synthesis of a functional gene product. These products are often proteins, but in non-protein coding genes such as rRNA genes or tRNA genes, the product is a functional RNA. Several steps in the Gene expression process may be modulated, including the transcription, RNA splicing, translation, and post-translational modification of a protein. Gene regulation gives the cell control over structure and function, and is the basis for cellular differentiation, morphogenesis and the versatility and adaptability of any organism.
Genetic	Genetics is the study of how living things receive common traits from previous generations. These traits are described by the Genetic information carried by a molecule called DNA. The instructions for constructing and operating an organism are contained in the organism's DNA. Every living thing on earth has DNA in its cells. Genes are the hereditary components of DNA that occupy spots on chromosomes and determine characteristics in an organism.
Genetic code	The Genetic code is the set of rules by which information encoded in genetic material (DNA or RNA sequences) is translated into proteins (amino acid sequences) by living cells. The code defines a mapping between tri-nucleotide sequences, called codons, and amino acids. A triplet codon in a nucleic acid sequence usually specifies a single amino acid (though in some cases the same codon triplet in different locations can code unambiguously for two different amino acids, the correct choice at each location being determined by context.)
Inborn errors of metabolism	Inborn errors of metabolism comprise a large class of genetic diseases involving disorders of metabolism. The majority are due to defects of single genes that code for enzymes that facilitate conversion of various substances (substrates) into others (products.) In most of the disorders, problems arise due to accumulation of substances which are toxic or interfere with normal function, or to the effects of reduced ability to synthesize essential compounds.
Information	Information as a concept has a diversity of meanings, from everyday usage to technical settings. Generally speaking, the concept of Information is closely related to notions of constraint, communication, control, data, form, instruction, knowledge, meaning, mental stimulus, pattern, perception, and representation.

	According to the Oxford English Dictionary, the first known historical meaning of the word Information in English was the act of informing, or giving form or shape to the mind, as in education, instruction, or training.
Translation	Translation is the first stage of protein biosynthesis (part of the overall process of gene expression.) Translation is the production of proteins by decoding mRNA produced in transcription. Translation occurs in the cytoplasm where the ribosomes are located.
Promoter	In genetics, a Promoter is a region of DNA that facilitates the transcription of a particular gene. Promoter s are typically located near the genes they regulate, on the same strand and upstream (towards the 5' region of the sense strand.) In order for transcription to take place, the enzyme that synthesizes RNA, known as RNA polymerase, must attach to the DNA near a gene.
Intron	An Intron is a DNA region within a gene that is not translated into protein. These non-coding sections are transcribed to precursor mRNA (pre-mRNA) and some other RNAs (such as long noncoding RNAs), and subsequently removed by a process called splicing during the processing to mature RNA. After Intron splicing (ie. removal), the mRNA consists only of exon derived sequences, which are translated into a protein.
Polyribosomes	Polyribosomes are a cluster of ribosomes, bound to a mRNA molecule, first discovered and characterized by Jonathan Warner, Paul Knopf, and Alex Rich in 1963. Polyribosomes read one strand of mRNA simultaneously, helping to synthesize the same protein at different spots on the mRNA, mRNA being the 'messenger' in the process of protein synthesis. They may appear as clusters, linear arrays, or rosettes in routine: this is aided by the fact that mRNA is able to be twisted into a circular formation, creating a cycle of rapid ribosome recycling, and utilization of ribosomes.
Mutation	In biology, mutation s are changes to the nucleotide sequence of the genetic material of an organism. mutation s can be caused by copying errors in the genetic material during cell division, by exposure to ultraviolet or ionizing radiation, chemical mutagens, or viruses, or can be induced by the organism itself, by cellular processes such as hyper mutation . In multicellular organisms with dedicated reproductive cells, mutation s can be subdivided into germ line mutation s, which can be passed on to descendants through the reproductive cells, and somatic mutation s, which involve cells outside the dedicated reproductive group and which are not usually transmitted to descendants.
Transposons	Transposons are sequences of DNA that can move around to different positions within the genome of a single cell, a process called transposition. In the process, they can cause mutations and change the amount of DNA in the genome. Transposons were also once called 'jumping genes', and are examples of mobile genetic elements.
Enzymes	Enzymes are biomolecules that catalyze (i.e., increase the rates of) chemical reactions. Nearly all known Enzymes are proteins. However, certain RNA molecules can be effective biocatalysts too.
Genetic engineering	Genetic engineering, recombinant DNA technology, genetic modification/manipulation (GM) and gene splicing are terms that apply to the direct manipulation of an organism's genes. Genetic engineering is different from traditional breeding, where the organism's genes are manipulated indirectly. Genetic engineering uses the techniques of molecular cloning and transformation to alter the structure and characteristics of genes directly.

Plasmid

A Plasmid is an extra-chromosomal DNA molecule separate from the chromosomal DNA which is capable of replicating independently of the chromosomal DNA. In many cases, it is circular and double-stranded. Plasmid s usually occur naturally in bacteria, but are sometimes found in eukaryotic organisms (e.g., the 2-micrometre-ring in Saccharomyces cerevisiae.)

Recombinant DNA

Recombinant DNA is a form of DNA that does not exist naturally, which is created by combining DNA sequences that would not normally occur together. In terms of genetic modification, Recombinant DNA is introduced through the addition of relevant DNA into an existing organismal DNA, such as the plasmids of bacteria, to code for or alter different traits for a specific purpose, such as antibiotic resistance. It differs from genetic recombination, in that it does not occur through processes within the cell, but is engineered.

Restriction enzyme

A restriction enzyme is an enzyme that cuts double-stranded or single stranded DNA at specific recognition nucleotide sequences known as restriction sites. Such enzymes, found in bacteria and archaea, are thought to have evolved to provide a defense mechanism against invading viruses. Inside a bacterial host, the restriction enzymes selectively cut up foreign DNA in a process called restriction; host DNA is methylated by a modification enzyme (a methylase) to protect it from the restriction enzyme's activity.

Cloning

Cloning in biology is the process of producing populations of genetically-identical individuals that occurs in nature when organisms such as bacteria, insects or plants reproduce asexually. Cloning in biotechnology refers to processes used to create copies of DNA fragments (molecular Cloning), cells (cell Cloning), or organisms. More generally, the term refers to the production of multiple copies of a product such as digital media or software.

Polymerase chain reaction

In molecular biology, the Polymerase chain reaction is a technique to amplify a single or few copies of a piece of DNA across several orders of magnitude, generating millions or more copies of a particular DNA sequence. The method relies on thermal cycling, consisting of cycles of repeated heating and cooling of the reaction for DNA melting and enzymatic replication of the DNA. Primers (short DNA fragments) containing sequences complementary to the target region along with a DNA polymerase (after which the method is named) are key components to enable selective and repeated amplification. As Polymerase chain reaction progresses, the DNA generated is itself used as a template for replication, setting in motion a chain reaction in which the DNA template is exponentially amplified.

Primer

A Primer is a strand of nucleic acid that serves as a starting point for DNA replication. They are required because the enzymes that catalyze replication, DNA polymerases, can only add new nucleotides to an existing strand of DNA. The polymerase starts replication at the 3'-end of the Primer, and copies the opposite strand.

In most cases of natural DNA replication, the Primer for DNA synthesis and replication is a short strand of RNA (which can be made de novo.)

DNA sequence

A DNA sequence or genetic sequence is a succession of letters representing the primary structure of a real or hypothetical DNA molecule or strand, with the capacity to carry information as described by the central dogma of molecular biology.

The possible letters are A, C, G, and T, representing the four nucleotide bases of a DNA strand -- adenine, cytosine, guanine, thymine -- covalently linked to a phosphodiester backbone. In the typical case, the sequences are printed abutting one another without gaps, as in the sequence AAAGTCTGAC, read left to right in the 5' to 3' direction.

Genome	In classical genetics, the Genome of a diploid organism including eukarya refers to a full set of chromosomes or genes in a gamete; thereby, a regular somatic cell contains two full sets of Genome s. In haploid organisms, including bacteria, archaea, viruses, and mitochondria, a cell contains only a single set of the Genome usually in a single circular or contiguous linear DNA (or RNA for retroviruses.) In modern molecular biology the Genome of an organism is its hereditary information encoded in DNA (or, for retroviruses, RNA.)
Genome Project	Genome project s are scientific endeavours that ultimately aim to determine the complete genome sequence of an organism (be it an animal, a plant, a fungus, a bacterium, an archaean, a protist or a virus.) The genome sequence for any organism requires the DNA sequences for each of the chromosomes in an organism to be determined. For bacteria, which usually have just one chromosome, a Genome project will aim to map the sequence of that chromosome.
Human	A Human is a member of a species of bipedal primates in the family Hominidae . DNA and fossil evidence indicates that modern Human s originated in east Africa about 200,000 years ago. When compared to other animals and primates, Human s have a highly developed brain, capable of abstract reasoning, language, introspection and problem solving.
Human Genome	The Human genome is the genome of Homo sapiens, which is stored on 23 chromosome pairs. Twenty-two of these are autosomal chromosome pairs, while the remaining pair is sex-determining. The haploid Human genome occupies a total of just over 3 billion DNA base pairs.
Human Genome Project	The Human Genome Project was an international scientific research project with a primary goal to determine the sequence of chemical base pairs which make up DNA and to identify and map the approximately 20,000-25,000 genes of the human genome from both a physical and functional standpoint. The project began in 1990 initially headed by James D. Watson at the U.S. National Institutes of Health. A working draft of the genome was released in 2000 and a complete one in 2003, with further analysis still being published.
Genomic	Genomic s is the study of the genomes of organisms. The field includes intensive efforts to determine the entire DNA sequence of organisms and fine-scale genetic mapping efforts. The field also includes studies of intra Genomic phenomena such as heterosis, epistasis, pleiotropy and other interactions between loci and alleles within the genome.
Sequencing	In genetics and biochemistry, Sequencing means to determine the primary structure (or primary sequence) of an unbranched biopolymer. Sequencing results in a symbolic linear depiction known as a sequence which succinctly summarizes much of the atomic-level structure of the sequenced molecule. DNA Sequencing is the process of determining the nucleotide order of a given DNA fragment.
Bioinformatics	Bioinformatics is the application of information technology to the field of molecular biology. The term Bioinformatics was coined by Paulien Hogeweg in 1978 for the study of informatic processes in biotic systems. Bioinformatics now entails the creation and advancement of databases, algorithms, computational and statistical techniques, and theory to solve formal and practical problems arising from the management and analysis of biological data.
Chimpanzee	Chimpanzee sometimes colloquially chimp, is the common name for the two extant species of ape in the genus Pan where the Congo River forms the boundary between the native habitat of the two species:

· Common Chimpanzee Pan troglodytes: the better known Chimpanzee lives primarily in West and Central Africa.
· Bonobo, Pan paniscus: also known as the 'Pygmy Chimpanzee , this species is found in the forests of the Democratic Republic of the Congo.
Chimpanzee s are members of the Hominidae family, along with gorillas, humans, and orangutans. Chimpanzee are thought to have split from human evolution about 6 million years ago and thus the two Chimpanzee species are the closest living relatives to humans, all being members of the Hominini tribe (along with extinct species of Hominina subtribe.) Chimpanzee s are the only known members of the Panina subtribe. The two Pan species split only about one million years ago.

Proteome

The Proteome is the entire complement of proteins expressed by a genome, cell, tissue or organism. More specifically, it is the set of expressed proteins at a given time under defined conditions. The term is a portmanteau of proteins and genome.

Proteomics

Proteomics is the large-scale study of proteins, particularly their structures and functions. Proteins are vital parts of living organisms, as they are the main components of the physiological metabolic pathways of cells. The term 'Proteomics' was first coined in 1997 to make an analogy with genomics, the study of the genes.

Crassulacean acid metabolism	Crassulacean acid metabolism is an elaborate carbon fixation pathway in some plants. These plants fix carbon dioxide during the night, storing it as the four carbon acid malate. The CO_2 is released during the day, where it is concentrated around the enzyme RuBisCO, increasing the efficiency of photosynthesis.
Cancer	Cancer is a genetic disorder in which the normal control of cell growth is lost. Cancer genetics is now one of the fastest expanding medical specialties. At the molecular level, Cancer is caused by mutation(s) in DNA, which result in aberrant cell proliferation.
Genome	In classical genetics, the Genome of a diploid organism including eukarya refers to a full set of chromosomes or genes in a gamete; thereby, a regular somatic cell contains two full sets of Genome s. In haploid organisms, including bacteria, archaea, viruses, and mitochondria, a cell contains only a single set of the Genome usually in a single circular or contiguous linear DNA (or RNA for retroviruses.) In modern molecular biology the Genome of an organism is its hereditary information encoded in DNA (or, for retroviruses, RNA.)
Genome Project	Genome project s are scientific endeavours that ultimately aim to determine the complete genome sequence of an organism (be it an animal, a plant, a fungus, a bacterium, an archaean, a protist or a virus.) The genome sequence for any organism requires the DNA sequences for each of the chromosomes in an organism to be determined. For bacteria, which usually have just one chromosome, a Genome project will aim to map the sequence of that chromosome.
Human	A Human is a member of a species of bipedal primates in the family Hominidae . DNA and fossil evidence indicates that modern Human s originated in east Africa about 200,000 years ago. When compared to other animals and primates, Human s have a highly developed brain, capable of abstract reasoning, language, introspection and problem solving.
Human Genome	The Human genome is the genome of Homo sapiens, which is stored on 23 chromosome pairs. Twenty-two of these are autosomal chromosome pairs, while the remaining pair is sex-determining. The haploid Human genome occupies a total of just over 3 billion DNA base pairs.
Human Genome Project	The Human Genome Project was an international scientific research project with a primary goal to determine the sequence of chemical base pairs which make up DNA and to identify and map the approximately 20,000-25,000 genes of the human genome from both a physical and functional standpoint. The project began in 1990 initially headed by James D. Watson at the U.S. National Institutes of Health. A working draft of the genome was released in 2000 and a complete one in 2003, with further analysis still being published.
Cloning	Cloning in biology is the process of producing populations of genetically-identical individuals that occurs in nature when organisms such as bacteria, insects or plants reproduce asexually. Cloning in biotechnology refers to processes used to create copies of DNA fragments (molecular Cloning), cells (cell Cloning), or organisms. More generally, the term refers to the production of multiple copies of a product such as digital media or software.
Oncogene	An Oncogene is a gene that, when mutated or expressed at high levels, helps turn a normal cell into a cancer cell. Many cells normally undergo a programmed form of death (apoptosis.) Activated Oncogene s can cause those cells to survive and proliferate instead.

Adult stem cells

Adult stem cells are undifferentiated cells, found throughout the body after embryonic development, that multiply by cell division to replenish dying cells and regenerate damaged tissues. Also known as somatic stem cells , they can be found in juvenile as well as adult animals and humans.

Scientific interest in Adult stem cells has centered on their ability to divide or self-renew indefinitely, and generate all the cell types of the organ from which they originate, potentially regenerating the entire organ from a few cells.

Embryonic stem cells

Embryonic stem cells are stem cells derived from the inner cell mass of an early stage embryo known as a blastocyst. Human embryos reach the blastocyst stage 4-5 days post fertilization, at which time they consist of 50-150 cells.

Embryonic Stem (ES) cells are pluripotent.

Stem

A stem is one of two main structural axes of a vascular plant. The stem is normally divided into nodes and internodes, the nodes hold buds which grow into one or more leaves, inflorescence (flowers), cones or other stems etc. The internodes act as spaces that distance one node from another.

Escherichia

Escherichia is a genus of Gram-negative, non-spore forming, facultatively anaerobic, rod-shaped bacteria from the family Enterobacteriaceae. Inhabitants of the gastrointestinal tracts of warm-blooded animals, Escherichia species provide a portion of the microbially-derived vitamin K for their host.

While many Escherichia are harmless commensals, particular strains of some species are human pathogens, and are known as the most common cause of urinary tract infections, significant sources of gastrointestinal disease, ranging from simple diarrhea to dysentery-like conditions, as well as a wide-range of other pathogenic states.

Escherichia coli

Escherichia coli , is a Gram negative bacterium that is commonly found in the lower intestine of warm-blooded organisms . Most E. coli strains are harmless, but some, such as serotype O157:H7, can cause serious food poisoning in humans, and are occasionally responsible for costly product recalls. The harmless strains are part of the normal flora of the gut, and can benefit their hosts by producing vitamin K_2, or by preventing the establishment of pathogenic bacteria within the intestine.

Gene expression

Gene expression is the process by which information from a gene is used in the synthesis of a functional gene product. These products are often proteins, but in non-protein coding genes such as rRNA genes or tRNA genes, the product is a functional RNA.

Several steps in the Gene expression process may be modulated, including the transcription, RNA splicing, translation, and post-translational modification of a protein. Gene regulation gives the cell control over structure and function, and is the basis for cellular differentiation, morphogenesis and the versatility and adaptability of any organism.

Operon

An Operon is a functioning unit of key nucleotide sequences of DNA including an operator, a common promoter, and one or more structural genes, which is controlled as a unit to produce messenger RNA (mRNA), in the process of transcription by an RNA polymerase. A typical Operon.

The term "Operon" was first proposed in a short paper in the Proceedings of the French Academy of Science in 1960. From this paper, the so-called general theory of the Operon was developed.

Promoter

In genetics, a Promoter is a region of DNA that facilitates the transcription of a particular gene. Promoter s are typically located near the genes they regulate, on the same strand and upstream (towards the 5' region of the sense strand.)

In order for transcription to take place, the enzyme that synthesizes RNA, known as RNA polymerase, must attach to the DNA near a gene.

Repressor

A Repressor is a DNA-binding protein that regulates the expression of one or more genes by decreasing the rate of transcription. This blocking of expression is called repression.

Repressor proteins are coded for by regulator genes.

Barr body

In those species (including humans) in which sex is determined by the presence of the Y or W chromosome rather than the diploidy of the X or Z, a Barr body is the inactive X chromosome in a female cell 2003), rendered inactive in a process called Lyonization. The Lyon hypothesis states that in cells with multiple X chromosomes, all but one are inactivated during mammalian embryogenesis (Lyon, 1961.) This happens early in embryonic development at random in mammals, (Brown, 1997) except in marsupials and in some extra-embryonic tissues of some placental mammals, in which the father's X chromosome is always deactivated (Lee, 2003.)

MyoD

MyoD is a protein with a key role in regulating muscle differentiation. MyoD belongs to a family of proteins known as myogenic regulatory factors . These bHLH (basic helix loop helix) transcription factors act sequentially in myogenic differentiation.

Activator

An Activator is a DNA-binding protein that regulates one or more genes by increasing the rate of transcription. The Activator may increase transcription by virtue of a connected domain which assists in the formation of the RNA polymerase holoenzyme, or may operate through a coactivator. A coactivator binds the DNA-binding Activator and contains the domain assisting holoenzyme formation.

Enhancer

In genetics, an Enhancer is a short region of DNA that can be bound with proteins (namely, the trans-acting factors, much like a set of transcription factors) to enhance transcription levels of genes (hence the name) in a gene cluster. Furthermore, an Enhancer is a cis-acting DNA sequence(s) which can increase transcription of genes. An Enhancer does not need to be particularly close to the genes it acts on, and need not be located on the same chromosome.

Euchromatin

Euchromatin is a lightly packed form of chromatin that is rich in gene concentration, and is often (but not always) under active transcription. Unlike heterochromatin, it is found in both eukaryotes and prokaryotes. Euchromatin comprises the most active portion of the genome within the cell nucleus.

Heterochromatin

Heterochromatin is a tightly packed form of DNA. Its major characteristic is that transcription is limited. As such, it is a means to control gene expression, through regulation of the transcription initiation.

Chromatin is found in two varieties: euchromatin and Heterochromatin.

Alagille syndrome

Alagille syndrome is a genetic disorder that affects the liver, heart, and other systems of the body. Problems associated with the disorder generally become evident in infancy or early childhood. The disorder is inherited in an autosomal dominant pattern, and the estimated prevalence of Alagille syndrome is 1 in every 100,000 live births.

Proteasome

Proteasomes are large protein complexes inside all eukaryotes and archaea, as well as in some bacteria. In eukaryotes, they are located in the nucleus and the cytoplasm. The main function of the Proteasome is to degrade unneeded or damaged proteins by proteolysis, a chemical reaction that breaks peptide bonds.

Translation

Translation is the first stage of protein biosynthesis (part of the overall process of gene expression.) Translation is the production of proteins by decoding mRNA produced in transcription. Translation occurs in the cytoplasm where the ribosomes are located.

Shigella

Shigella is a genus of Gram-negative, non-spore forming rod-shaped bacteria closely related to Escherichia coli and Salmonella. The causative agent of human shigellosis, Shigella cause disease in primates, but not in other mammals. It is only naturally found in humans and apes.

Shigella dysenteriae

Shigella dysenteriae is a species of the rod-shaped bacterial genus Shigella. Shigella can cause shigellosis (bacillary dysentery.) Shigellae are Gram-negative, non-spore-forming, facultatively anaerobic, non-motile bacteria.

Signal transduction

In biology, 'Signal transduction' refers to any process by which a cell converts one kind of signal or stimulus into another. Most processes of Signal transduction involve ordered sequences of biochemical reactions inside the cell, which are carried out by enzymes and activated by second messengers, resulting in a Signal transduction pathway. Such processes are usually rapid, lasting on the order of milliseconds in the case of ion flux, or minutes for the activation of protein- and lipid-mediated kinase cascades, but some can take hours, and even days (as is the case with gene expression), to complete.

Horse murders

The Horse murders scandal was a form of insurance fraud in the United States in which expensive horses, many of them show jumpers, were insured against death, accident and then killed to collect the insurance money. It is not known how many horses were murdered between the mid 1970s and the mid-1990s, when a Federal Bureau of Investigation (FBI) investigation brought the horse killings to light, but the number is thought to be well over 50, and may have been as high as 100. In addition, in 1977, the heiress Helen Brach disappeared and was presumed by law enforcement agents to have been murdered by the perpetrators of these crimes, because she threatened to report their criminal activity to authorities; continuing investigations into Brach's death began to uncover the insurance fraud in the 1990s.

Growth factor

A Growth factor is a naturally occurring substance capable of stimulating cellular growth, proliferation and cellular differentiation. Usually it is a protein or a steroid hormone. Growth factor s are important for regulating a variety of cellular processes.

Tumor

A tumor or tumour is the name for a swelling or lesion formed by an abnormal growth of cells (termed neoplastic.) tumor is not synonymous with cancer. A tumor can be benign, pre-malignant or malignant, whereas cancer is by definition malignant.

Tumor suppressor gene

A Tumor suppressor gene is a gene that protects a cell from one step on the path to cancer. When this gene is mutated to cause a loss or reduction in its function, the cell can progress to cancer, usually in combination with other genetic changes.

Unlike oncogenes, Tumor suppressor gene s generally follow the 'two-hit hypothesis', which implies that both alleles that code for a particular gene must be affected before an effect is manifested.

Chromosome	A Chromosome is an organized structure of DNA and protein that is found in cells. It is a single piece of coiled DNA containing many genes, regulatory elements and other nucleotide sequences. Chromosome s also contain DNA-bound proteins, which serve to package the DNA and control its functions.
Protein	Protein s are organic compounds made of amino acids arranged in a linear chain. The amino acids in a polymer chain are joined together by the peptide bonds between the carboxyl and amino groups of adjacent amino acid residues. The sequence of amino acids in a protein is defined by the sequence of a gene, which is encoded in the genetic code.
Retinoblastoma protein	The Retinoblastoma protein is a tumor suppressor protein that is dysfunctional in many types of cancer. One highly studied function of pRb is to prevent excessive cell growth by inhibiting cell cycle progression until a cell is ready to divide. pRb belongs to the pocket protein family, whose members have a pocket for the functional binding of other proteins.
Telomerase	Telomerase is an enzyme that adds specific DNA sequence repeats ('TTAGGG' in all vertebrates) to the 3' end of DNA strands in the telomere regions, which are found at the ends of eukaryotic chromosomes. The telomeres contain condensed DNA material, giving stability to the chromosomes. The enzyme is a reverse transcriptase that carries its own RNA molecule, which is used as a template when it elongates telomeres, which are shortened after each replication cycle.
Telomere	A Telomere is a region of repetitive DNA at the end of chromosomes, which protects the end of the chromosome from destruction. A Russian theorist Alexei Olovnikov was the first to recognize the problem of how chromosomes could replicate right to the tip, as such was impossible with replication in a 3' to 5' direction. To solve this and to accommodate Leonard Hayflick's idea of limited somatic cell division, Olovnikov suggested that DNA sequences would be lost in every replicative phase until they reached a critical level, at which point cell division would stop.

Genetic

Genetics is the study of how living things receive common traits from previous generations. These traits are described by the Genetic information carried by a molecule called DNA. The instructions for constructing and operating an organism are contained in the organism's DNA. Every living thing on earth has DNA in its cells. Genes are the hereditary components of DNA that occupy spots on chromosomes and determine characteristics in an organism.

Genetic counseling

Genetic counseling is the process by which patients or relatives, at risk of an inherited disorder, are advised of the consequences and nature of the disorder, the probability of developing or transmitting it, and the options open to them in management and family planning in order to prevent, avoid or ameliorate it. This complex process can be seen from diagnostic (the actual estimation of risk) and supportive aspects.

A genetic counselor is a medical genetics expert with a master of science degree.

Karyotype

A Karyotype is the characteristic chromosome complement of a eukaryote species. The preparation and study of Karyotype s is part of cytogenetics. Karyogram of human male using Giemsa staining.

The basic number of chromosomes in the somatic cells of an individual or a species is called the somatic number and is designated 2n.

DNA

Deoxyribonucleic acid (DNA) is a nucleic acid that contains the genetic instructions used in the development and functioning of all known living organisms and some viruses. The main role of DNA molecules is the long-term storage of information. DNA is often compared to a set of blueprints or a recipe, or a code, since it contains the instructions needed to construct other components of cells, such as proteins and RNA molecules.

Technology

Technology is a broad concept that deals with an animal species' ethology or behavior of usage and of knowledge of tools and crafts, and how it affects the animal species' ability to control and adapt to its environment. Technology is a term with origins in the Greek 'technologia', 'τεχνολογΐ¯α' -- 'techne', 'τΐχνη' and 'logia', 'λογΐ¯α' ('saying'.) However, a strict definition is elusive; 'Technology' can refer to material objects of use to humanity, such as machines, hardware or utensils, but can also encompass broader themes, including systems, methods of organization, and techniques.

Cri du chat

Cri du chat syndrome 5p minus syndrome or Lejeune's syndrome, is a rare genetic disorder due to a missing part of chromosome 5. Its name is a French term referring to the characteristic cat-like cry of affected children. It was first described by Jérôme Lejeune in 1963.

Decomposers

Decomposers are organisms that consume dead or decaying organisms, and, in doing so, carry out the natural process of decomposition. Like herbivores and predators, Decomposers are heterotrophic, meaning that they use organic substrates to get their energy, carbon and nutrients for growth and development. Decomposers use deceased organisms and non-living organic compounds as their food source.

Deletion

In genetics, a Deletion is a mutation in which a part of a chromosome or a sequence of DNA is missing. Deletion is the loss of genetic material. Any number of nucleotides can be deleted, from a single base to an entire piece of chromosome.

Inversion

An inversion is a chromosome rearrangement in which a segment of a chromosome is reversed end to end. An inversion occurs when a single chromosome undergoes breakage and rearrangement within itself. Inversions are of two types: paracentric and pericentric.

Mutation

In biology, mutation s are changes to the nucleotide sequence of the genetic material of an organism. mutation s can be caused by copying errors in the genetic material during cell division, by exposure to ultraviolet or ionizing radiation, chemical mutagens, or viruses, or can be induced by the organism itself, by cellular processes such as hyper mutation . In multicellular organisms with dedicated reproductive cells, mutation s can be subdivided into germ line mutation s, which can be passed on to descendants through the reproductive cells, and somatic mutation s, which involve cells outside the dedicated reproductive group and which are not usually transmitted to descendants.

Alagille syndrome

Alagille syndrome is a genetic disorder that affects the liver, heart, and other systems of the body. Problems associated with the disorder generally become evident in infancy or early childhood. The disorder is inherited in an autosomal dominant pattern, and the estimated prevalence of Alagille syndrome is 1 in every 100,000 live births.

Cell

The Cell is the structural and functional unit of all known living organisms. It is the smallest unit of an organism that is classified as living, and is often called the building block of life. Some organisms, such as most bacteria, are unicellular (consist of a single Cell.)

Family

In biological classification, Family is

· a taxonomic rank. Other well-known ranks are life, domain, kingdom, phylum, class, order, genus, and species, with Family fitting between order and genus. As for the other well-known ranks, there is the option of an immediately lower rank, indicated by the prefix sub-: subfamily .

· a taxonomic unit, a taxon, in that rank. In that case the plural is families

Example: 'Walnuts and Hickories belong to the Walnut Family.

What does and does not belong to each Family is determined by a taxonomist. Similarly for the question if a particular Family should be recognized at all. Often there is no exact agreement, with different taxonomists each taking a different position.

Genetic disorder

A Genetic disorder is an illness caused by abnormalities in genes or chromosomes. While some diseases, such as cancer, are due in part to a Genetic disorder s, they can also be caused by environmental factors. Most disorders are quite rare and affect one person in every several thousands or millions.

Trait

A trait is a distinct variant of a phenotypic character of an organism that may be inherited, environmentally determined or somewhere in between. For example, eye color is a character or abstraction of an attribute, while blue, brown and hazel are traits.

A trait may be any single feature or quantifiable measurement of an organism.

Alkaptonuria

Alkaptonuria is a rare inherited genetic disorder of phenylalanine and tyrosine metabolism. This is an autosomal recessive condition that is due to a defect in the enzyme homogentisate 1,2-dioxygenase (EC 1.13.11.5), which participates in the degradation of tyrosine. As a result, a toxic tyrosine byproduct called homogentisic acid (or alkapton) accumulates in the blood and is excreted in urine in large amounts(hence -uria.)

Cystic fibrosis

Cystic fibrosis is a genetic disorder known to be an inherited disease of the secretory glands, including the glands that make mucus and sweat.

	The hallmarks of Cystic fibrosis are salty tasting skin, normal appetite but poor growth and poor weight gain, excess mucus production, and coughing/shortness of breath. Males can be infertile due to the condition congenital bilateral absence of the vas deferens.
Methemoglobinemia	Methemoglobinemia is a disorder characterized by the presence of a higher than normal level of methemoglobin (metHb) in the blood. Methemoglobin is a form of hemoglobin that does not bind oxygen. When its concentration is elevated in red blood cells, tissue hypoxia can occur.
Shigella	Shigella is a genus of Gram-negative, non-spore forming rod-shaped bacteria closely related to Escherichia coli and Salmonella. The causative agent of human shigellosis, Shigella cause disease in primates, but not in other mammals. It is only naturally found in humans and apes.
Shigella dysenteriae	Shigella dysenteriae is a species of the rod-shaped bacterial genus Shigella. Shigella can cause shigellosis (bacillary dysentery.) Shigellae are Gram-negative, non-spore-forming, facultatively anaerobic, non-motile bacteria.
Horse murders	The Horse murders scandal was a form of insurance fraud in the United States in which expensive horses, many of them show jumpers, were insured against death, accident and then killed to collect the insurance money. It is not known how many horses were murdered between the mid 1970s and the mid-1990s, when a Federal Bureau of Investigation (FBI) investigation brought the horse killings to light, but the number is thought to be well over 50, and may have been as high as 100. In addition, in 1977, the heiress Helen Brach disappeared and was presumed by law enforcement agents to have been murdered by the perpetrators of these crimes, because she threatened to report their criminal activity to authorities; continuing investigations into Brach's death began to uncover the insurance fraud in the 1990s.
DNA microarray	A DNA microarray is a multiplex technology used in molecular biology and in medicine. It consists of an arrayed series of thousands of microscopic spots of DNA oligonucleotides, called features, each containing picomoles of a specific DNA sequence. This can be a short section of a gene or other DNA element that are used as probes to hybridize a cDNA or cRNA sample (called target) under high-stringency conditions.
Genetic marker	A Genetic marker is a gene or DNA sequence with a known location on a chromosome and associated with a particular gene or trait. It can be described as a variation, which may arise due to mutation or alteration in the genomic loci, that can be observed. A Genetic marker may be a short DNA sequence, such as a sequence surrounding a single base-pair change (single nucleotide polymorphism, SNP), or a long one, like minisatellites.
Edward	The Edward mango is a monoembryonic mango cultivar grown predominantly in Florida. It is considered by many to be among the finest tasting mangoes in the world; however, its poor yields have restrained the Edward from developing into a commercially significant variety. The Edward was first propagated in the 1920s by Edward Simmonds of the Plant Introduction Garden in Miami, Florida and is believed to be a hybrid cross of Haden and Carabao mango cultivars.
Fetus	A Fetus (also spelled foetus or fÅ"tus) is a developing mammal or other viviparous vertebrate, after the embryonic stage and before birth. The plural is Fetus es. In humans, the fetal stage of prenatal development starts at the beginning of the 11th week in gestational age (the 9th week after fertilization.)

Escherichia

Escherichia is a genus of Gram-negative, non-spore forming, facultatively anaerobic, rod-shaped bacteria from the family Enterobacteriaceae. Inhabitants of the gastrointestinal tracts of warm-blooded animals, Escherichia species provide a portion of the microbially-derived vitamin K for their host.

While many Escherichia are harmless commensals, particular strains of some species are human pathogens, and are known as the most common cause of urinary tract infections, significant sources of gastrointestinal disease, ranging from simple diarrhea to dysentery-like conditions, as well as a wide-range of other pathogenic states.

Escherichia coli

Escherichia coli , is a Gram negative bacterium that is commonly found in the lower intestine of warm-blooded organisms . Most E. coli strains are harmless, but some, such as serotype O157:H7, can cause serious food poisoning in humans, and are occasionally responsible for costly product recalls. The harmless strains are part of the normal flora of the gut, and can benefit their hosts by producing vitamin K_2, or by preventing the establishment of pathogenic bacteria within the intestine.

Egg

Phasmatodea, like a ghost, resembles twigs leaves and sometimes resembles lichen. The Egg is oval with 5 parts making the Egg, the capitulum-small nob on end of Egg, operculum-the ring holding the capitulum and is in between the shell and the capitulum. The shell which is the larger oval shaped part to the Egg has the nymph and the sheet separating the yolk and the shell to which the nymph emerges from the connected oper and capitulum, sometimes when the nymph emerges, say the tropidoderus childrenii when the nymph hatches with the Egg still attached to the hind legs causing leg dis-formation.

Eugenics

Eugenics is 'the study of, or belief in, the possibility of improving the qualities of the human species or a human population by such means as discouraging reproduction by persons having genetic defects or presumed to have inheritable undesirable traits (negative Eugenics) or encouraging reproduction by persons presumed to have inheritable desirable traits (positive Eugenics.)'

As a social movement Eugenics reached its height of popularity in the early decades of the 20th century. By the end of World War II Eugenics had been largely abandoned, though current trends in genetics have raised questions amongst critical academics concerning parallels between pre-war attitudes about Eugenics and current 'utilitarian' and social darwinistic theories. At its pre-war zenith, the movement often pursued pseudoscientific notions of racial supremacy and purity.

Ex vivo

Ex vivo means that which takes place outside an organism. In science, Ex vivo refers to experimentation or measurements done in or on living tissue in an artificial environment outside the organism with the minimum alteration of the natural conditions. The most common 'Ex vivo' procedures involve living cells or tissues taken from an organism and cultured in a laboratory apparatus, usually under sterile conditions and no alterations done for a few hours up to 24 hrs.

Familial hypercholesterolemia

Familial hypercholesterolemia is a genetic disorder characterized by high cholesterol levels, specifically very high low-density lipoprotein (LDL, 'bad cholesterol') levels, in the blood and early cardiovascular disease. Many patients have mutations in the LDLR gene that encodes the LDL receptor protein, which normally removes LDL from the circulation, or apolipoprotein B (ApoB), which is the part of LDL that binds with the receptor; mutations in other genes are rare. Patients who have one abnormal copy (are heterozygous) of the LDLR gene may have premature cardiovascular disease at the age of 30 to 40.

Gene	A Gene is the basic unit of heredity in a living organism. All living things depend on Gene s. Gene s hold the information to build and maintain their cells and pass Gene tic traits to offspring.
Gene therapy	Gene therapy is the insertion of genes into an individual's cells and tissues to treat a disease, such as a hereditary disease in which a deleterious mutant allele is replaced with a functional one. Although the technology is still in its infancy, it has been used with some success. Antisense therapy is not strictly a form of Gene therapy, but is a genetically-mediated therapy and is often considered together with other methods.
Severe combined immunodeficiency	Severe combined immunodeficiency is a genetic disorder in which both 'arms' of the adaptive immune system are crippled, due to a defect in one of several possible genes. Severe combined immunodeficiency D is a severe form of heritable immunodeficiency. It is also known as the 'bubble boy' disease because its victims are extremely vulnerable to infectious diseases.

Escherichia

Escherichia is a genus of Gram-negative, non-spore forming, facultatively anaerobic, rod-shaped bacteria from the family Enterobacteriaceae. Inhabitants of the gastrointestinal tracts of warm-blooded animals, Escherichia species provide a portion of the microbially-derived vitamin K for their host.

While many Escherichia are harmless commensals, particular strains of some species are human pathogens, and are known as the most common cause of urinary tract infections, significant sources of gastrointestinal disease, ranging from simple diarrhea to dysentery-like conditions, as well as a wide-range of other pathogenic states.

Escherichia coli

Escherichia coli , is a Gram negative bacterium that is commonly found in the lower intestine of warm-blooded organisms . Most E. coli strains are harmless, but some, such as serotype O157:H7, can cause serious food poisoning in humans, and are occasionally responsible for costly product recalls. The harmless strains are part of the normal flora of the gut, and can benefit their hosts by producing vitamin K_2, or by preventing the establishment of pathogenic bacteria within the intestine.

Charles Robert Darwin

Charles Robert Darwin FRS (12 February 1809 - 19 April 1882) was an English naturalist who realised and presented compelling evidence that all species of life have evolved over time from common ancestors, through the process he called natural selection. The fact that evolution occurs became accepted by the scientific community and much of the general public in his lifetime, while his theory of natural selection came to be widely seen as the primary explanation of the process of evolution in the 1930s, and now forms the basis of modern evolutionary theory. In modified form, Darwin's scientific discovery is the unifying theory of the life sciences, providing logical explanation for the diversity of life.

NADPH

Nicotinamide adenine dinucleotide phosphate ($NADP^+$, in older notation triphosphopyridine nucleotide, TPN) is used in anabolic reactions, such as lipid and nucleic acid synthesis, which require NADPH as a reducing agent.

NADPH is the reduced form of $NADP^+$, and $NADP^+$ is the oxidized form of NADPH. NADP+ differs from NAD+ by the presence in NADP+ of an additional phosphate group on the 2' position of the ribose ring that carries the adenine moiety. In chloroplasts, NADP is reduced by ferredoxin-NADP+ reductase in the last step of the electron chain of the light reactions of photosynthesis.

Artificial selection

Artificial selection describes intentional breeding for certain traits, or combination of traits. It was defined by Charles Darwin in contrast to natural selection, in which the differential reproduction of organisms with certain traits is attributed to improved survival or reproductive ability ('Darwinian fitness') and the creation of new organisms, us humans breed them for specific traits. Artificial selection can also be unintentional; it is thought that domestication of crops by early humans was largely unintentional.

Photosynthesis

Photosynthesis is a process that converts carbon dioxide into organic compounds, especially sugars, using the energy from sunlight. Photosynthesis occurs in plants, algae, and many species of Bacteria, but not in Archaea. Photosynthetic organisms are called photoautotrophs, since it allows them to create their own food.

Paleontology

Paleontology is the study of prehistoric life, including organisms' evolution and interactions with each other and their environments (their paleoecology.) As a 'historical science' it tries to explain causes rather than conduct experiments to observe effects. Paleontological observations have been documented as far back as the 5th century BC. The science became established in the 18th century as a result of Georges Cuvier's work on comparative anatomy, and developed rapidly in the 19th century.

Gene — A Gene is the basic unit of heredity in a living organism. All living things depend on Gene s. Gene s hold the information to build and maintain their cells and pass Gene tic traits to offspring.

Virus — A virus is a microscopic infectious agent that can reproduce only inside a host cell. virus es infect all types of organisms: from animals and plants, to bacteria and archaea. Since the initial discovery of tobacco mosaic virus by Martinus Beijerinck in 1898, more than 5,000 types of virus have been described in detail, although most types of virus remain undiscovered.

Anatomy — The anatomy of spiders is in some aspects similar, but also different from that of other arthropods. The following characteristics are common to all spiders: A body with two segments, eight legs, spinnerets, no chewing parts, no wings, and the presence of chelicerae, which spiders use to hold prey, and in most cases, inject venom. Spiders have non-compound eyes, with most species having eight; the spiders known as Haplogynae may have six or fewer, and certain cave-dwelling spiders may have none at all.

Embryo — An Embryo is a multicellular diploid eukaryote in its earliest stage of development, from the time of first cell division until birth, hatching, or germination. In humans, it is called an Embryo until about eight weeks after fertilization (i.e. ten weeks LMP), and from then it is instead called a fetus. 6 week old excised human Embryo /span>

The development of the Embryo is called Embryo genesis.

Scale

In most biological nomenclature, a Scale is a small rigid plate that grows out of an animal's skin to provide protection. In lepidopteran species, scales are plates on the surface of the insect wing, and provide coloration. Scales are quite common and have evolved multiple times with varying structure and function.

Ammagnostidae

Ammagnostidae is a family of trilobites in the superfamily Agnostoidea, which is part of the order Agnostida 2 or 3 segmented trilobites. It has four genera:

Ammagnostus Öpik, 1967

· Ammagnostus psammius Öpik, 1967 (Type)
· Ammagnostus bassus (Öpik, 1967) Guo and Luo, 1996
· Ammagnostus bella Guo and Luo, 1996
· Ammagnostus beltensis (Lochman, 1944) Robinson, 1988
· Ammagnostus cryptus
· Ammagnostus cylindratus Guo and Luo, 1996
· Ammagnostus duibianensis Lu and Lin, 1989
· Ammagnostus histus
· Ammagnostus hunanensis
· Ammagnostus integriceps Öpik, 1967
· Ammagnostus laiwuensis (Lorenz, 1906)
· Ammagnostus mitis Öpik, 1967
· Ammagnostus sinensis Peng, 1987
· Ammagnostus wangcunensis Peng and Robison

Hadragnostus Öpik, 1967

· Hadragnostus las Öpik, 1967 (Type)
· Hadragnostus edax Fortey and Rushton, 1976
· Hadragnostus helixensis Jago and Cooper, 2005
· Hadragnostus modestus (Lochman, 1944)

Kormagnostus Resser, 1938

· Kormagnostus sinplex Resser, 1938 (Type)
· Kormagnostus boltoni Westrop et al., 1996
· Kormagnostus copelandi Westrop et al., 1996
· Kormagnostus flati Pratt, 1992
· Kormagnostus inventus Shergold, 1982
· Kormagnostus minutus (Schrank, 1975)
· Kormagnostus seclusus (Walcott, 1884)

Proagnostus Butts, 1926

· Proagnostus bulbus Butts, 1926 (Type)
· Proagnostus centerensis Resser, 1938
· Proagnostus maryvillensis Resser

They occurred during the late Cambrian period to the Ordovician period.

Cell

The Cell is the structural and functional unit of all known living organisms. It is the smallest unit of an organism that is classified as living, and is often called the building block of life. Some organisms, such as most bacteria, are unicellular (consist of a single Cell.)

Population

In biology, a population is the collection of inter-breeding organisms of a particular species; in sociology, a collection of human beings. Individuals within a population share a factor may be reduced by statistical means, but such a generalization may be too vague to imply anything. Demography is used extensively in marketing, which relates to economic units, such as retailers, to potential customers.

Punnett square

The Punnett square is a diagram that is used to predict the outcome of a particular cross or breeding experiment. It is named after Reginald C. Punnett, who devised the approach, and is used by biologists to determine the probability of an offspring having a particular genotype. The Punnett square is a summary of every possible combination of one maternal allele with one paternal allele for each gene being studied in the cross.

Gene

A Gene is the basic unit of heredity in a living organism. All living things depend on Gene s. Gene s hold the information to build and maintain their cells and pass Gene tic traits to offspring.

Gene pool

In population genetics, a Gene pool is the complete set of unique alleles in a species or population.

A large Gene pool indicates extensive genetic diversity, which is associated with robust populations that can survive bouts of intense selection. Meanwhile, low genetic diversity can cause reduced biological fitness and an increased chance of extinction.

Genetic

Genetics is the study of how living things receive common traits from previous generations. These traits are described by the Genetic information carried by a molecule called DNA. The instructions for constructing and operating an organism are contained in the organism's DNA. Every living thing on earth has DNA in its cells. Genes are the hereditary components of DNA that occupy spots on chromosomes and determine characteristics in an organism.

Shigella

Shigella is a genus of Gram-negative, non-spore forming rod-shaped bacteria closely related to Escherichia coli and Salmonella. The causative agent of human shigellosis, Shigella cause disease in primates, but not in other mammals. It is only naturally found in humans and apes.

Shigella dysenteriae

Shigella dysenteriae is a species of the rod-shaped bacterial genus Shigella. Shigella can cause shigellosis (bacillary dysentery.) Shigellae are Gram-negative, non-spore-forming, facultatively anaerobic, non-motile bacteria.

Leaf

In botany, a Leaf is an above-ground plant organ specialized for photosynthesis. For this purpose, a Leaf is typically flat (laminar) and thin. There is continued debate about whether the flatness of leaves [[Natural selection | evolved] to expose the chloroplasts to more light or to increase the absorption of carbon dioxide.

Mutation

In biology, mutation s are changes to the nucleotide sequence of the genetic material of an organism. mutation s can be caused by copying errors in the genetic material during cell division, by exposure to ultraviolet or ionizing radiation, chemical mutagens, or viruses, or can be induced by the organism itself, by cellular processes such as hyper mutation . In multicellular organisms with dedicated reproductive cells, mutation s can be subdivided into germ line mutation s, which can be passed on to descendants through the reproductive cells, and somatic mutation s, which involve cells outside the dedicated reproductive group and which are not usually transmitted to descendants.

Nucleotide

Nucleotide s are molecules that, when joined together, make up the structural units of RNA and DNA. Additionally, Nucleotide s play central roles in metabolism. In that capacity, they serve as sources of chemical energy (adenosine triphosphate and guanosine triphosphate), participate in cellular signaling (cyclic guanosine monophosphate and cyclic adenosine monophosphate), and are incorporated into important cofactors of enzymatic reactions (coenzyme A, flavin adenine di Nucleotide , flavin mono Nucleotide , and nicotinamide adenine di Nucleotide phosphate.) Figure 1: Structural elements of the most common Nucleotide s Figure 2: Ribose structure indicating numbering of carbon atoms

A Nucleotide is composed of a nucleobase (nitrogenous base) and a five-carbon sugar (either ribose or 2'-deoxyribose), and one to three phosphate groups.

Polymorphism

Polymorphism in biophysics is the aspect of the behaviour of lipids that influences their long-range order, i.e. how they aggregate. This can be in the form of spheres of lipid molecules (micelles), pairs of layers that face one another (lamellar phase, observed in biological system as a lipid bilayer), a tubular arrangement (hexagonal), or various cubic phases (Fd3m, Im3m, Ia3m, Pn3m, and Pm3m being those discovered so far.) More complicated aggregations have also been observed, rhombohedral, tetragonal and orthorhombic phases have been observed.

Ascaris

Ascaris is a genus of parasitic nematode worms known as the giant intestinal roundworms. One species, A. suum, typically infects pigs, while another, A. lumbricoides, affects human populations, typically in sub-tropical and tropical areas with poor sanitation. A. lumbricoides is the largest intestinal roundworm and is the most common helminth infection of humans worldwide, an infection known as ascariasis.

Ascaris lumbricoides

Ascaris lumbricoides is the member of the Ascaris family responsible for the disease ascariasis.

It can reach a length of up to 35 cm.

Ascaris lumbricoides, or 'roundworm', infections in humans occur when an ingested infective egg releases a larval worm that penetrates the wall of the duodenum and enters the bloodstream.

Gene flow

In population genetics, gene flow is the transfer of alleles of genes from one population to another.

Migration into or out of a population may be responsible for a marked change in allele frequencies Immigration may also result in the addition of new genetic variants to the established gene pool of a particular species or population.

Genetic drift

Genetic drift or allelic drift is the change in the relative frequency with which a gene variant (allele) occurs in a population that results from the fact that alleles in offspring are a random sample of those in the parents, and because of the role of chance in determining whether a given individual survives and reproduces. A population's allele frequency is the fraction of the gene copies that share a particular form.

Genetic drift is one of several evolutionary processes which lead to changes in allele frequencies over time.

Mating

In biology, Mating is the pairing of opposite-sex or hermaphroditic organisms for copulation and, in social animals, also to raise their offspring. For animals, Mating methods include random Mating, disassortative Mating, assortative Mating, or a Mating pool.

In some birds, for example, it includes nest-building and feeding offspring.

Key Habitat Site

A Key Habitat Site is a Canadian Wildlife Service designation for an area that supports at least 1% of the country's population of any migratory bird species at any time. There may be overlap with areas designated as a migratory bird sanctuary or national wildlife area.

NADPH

Nicotinamide adenine dinucleotide phosphate ($NADP^+$, in older notation triphosphopyridine nucleotide, TPN) is used in anabolic reactions, such as lipid and nucleic acid synthesis, which require NADPH as a reducing agent.

NADPH is the reduced form of $NADP^+$, and $NADP^+$ is the oxidized form of NADPH. NADP+ differs from NAD+ by the presence in NADP+ of an additional phosphate group on the 2' position of the ribose ring that carries the adenine moiety.

In chloroplasts, NADP is reduced by ferredoxin-NADP+ reductase in the last step of the electron chain of the light reactions of photosynthesis.

Photosynthesis

Photosynthesis is a process that converts carbon dioxide into organic compounds, especially sugars, using the energy from sunlight. Photosynthesis occurs in plants, algae, and many species of Bacteria, but not in Archaea. Photosynthetic organisms are called photoautotrophs, since it allows them to create their own food.

Spirogyra

Spirogyra is a genus of filamentous green algae of the order Zygnematales and there are more than 400 species of Spirogyra in the world. Spirogyra measures approximately 10 to 100μm in width and may stretch centimeters long.

Spirogyra is unbranched with cylindrical cells connected end to end in long green filaments.

Horse murders

The Horse murders scandal was a form of insurance fraud in the United States in which expensive horses, many of them show jumpers, were insured against death, accident and then killed to collect the insurance money. It is not known how many horses were murdered between the mid 1970s and the mid-1990s, when a Federal Bureau of Investigation (FBI) investigation brought the horse killings to light, but the number is thought to be well over 50, and may have been as high as 100. In addition, in 1977, the heiress Helen Brach disappeared and was presumed by law enforcement agents to have been murdered by the perpetrators of these crimes, because she threatened to report their criminal activity to authorities; continuing investigations into Brach's death began to uncover the insurance fraud in the 1990s.

Maintenance

Maintenance of an organism is the collection of processes to stay alive, excluding production processes. The Dynamic Energy Budget theory delineates two classes

· Somatic Maintenance. This comprises the turnover of structural mass (mainly proteins), the Maintenance of concentration gradients of metabolites across membranes, activity

· Maturity Maintenance. This comprises the Maintenance of defence systems (such as the immune system), the preparation of the body for reproduction.

The theory assumes that maturity Maintenance costs can be reduced more easily during starvation than somatic Maintenance costs. Under extreme starvation conditions, somatic Maintenance costs are paid from structural mass, which causes shrinking.Some organism manage to switch to the turpor state under starvation conditions, and reduce their Maintenance costs.

Scale	In most biological nomenclature, a Scale is a small rigid plate that grows out of an animal's skin to provide protection. In lepidopteran species, scales are plates on the surface of the insect wing, and provide coloration. Scales are quite common and have evolved multiple times with varying structure and function.
Species	In biology, a Species is: · a taxonomic rank (the basic rank of biological classification) or · a unit at that rank There are many definitions of what kind of unit a Species is (or should be.) A common definition is that of a group of organisms capable of interbreeding and producing fertile offspring, and separated from other such groups with which interbreeding does not (normally) happen. Other definitions may focus on similarity of DNA or morphology. Some Species are further subdivided into sub Species , and here also there is no close agreement on the criteria to be used.
Horse murders	The Horse murders scandal was a form of insurance fraud in the United States in which expensive horses, many of them show jumpers, were insured against death, accident and then killed to collect the insurance money. It is not known how many horses were murdered between the mid 1970s and the mid-1990s, when a Federal Bureau of Investigation (FBI) investigation brought the horse killings to light, but the number is thought to be well over 50, and may have been as high as 100. In addition, in 1977, the heiress Helen Brach disappeared and was presumed by law enforcement agents to have been murdered by the perpetrators of these crimes, because she threatened to report their criminal activity to authorities; continuing investigations into Brach's death began to uncover the insurance fraud in the 1990s.
Gamete	A Gamete is a cell that fuses with another Gamete during fertilization (conception) in organisms that reproduce sexually. In species that produce two morphologically distinct types of Gamete s, and in which each individual produces only one type, a female is any individual that produces the larger type of Gamete -- called an ovum (or egg) -- and a male produces the smaller tadpole-like type -- called a sperm. This is an example of anisogamy or heterogamy, the condition wherein females and males produce Gamete s of different sizes (this is the case in humans; the human ovum is approximately 20 times larger than the human sperm cell.)
Hybrid	Hybrid has two meanings. The first meaning is the result of interbreeding between two animals or plants of different taxa. Hybrids between different subspecies within a species (such as between the Bengal tiger and Siberian tiger) are known as intra-specific hybrids.
Zygote	A zygote is a term in Developmental biology used to describe the first stage of a new unique organism when it consists of just a single cell. The term is also used more loosely to refer to the group of cells formed by the first few cell divisions, although this is properly referred to as a blastomere. A zygote is usually produced by a fertilization event between two haploid cells - an ovum from a female and a sperm cell from a male - which combine to form the single diploid cell.
Alagille syndrome	Alagille syndrome is a genetic disorder that affects the liver, heart, and other systems of the body. Problems associated with the disorder generally become evident in infancy or early childhood. The disorder is inherited in an autosomal dominant pattern, and the estimated prevalence of Alagille syndrome is 1 in every 100,000 live births.
Polyploidy	Polyploidy occurs in cells and organisms when there are more than two paired (homologous) sets of chromosomes. Known paleo Polyploidy in eukaryotes

Most organisms are normally diploid, meaning they have two sets of chromosomes -- one set inherited from each parent. Polyploidy may occur due to abnormal cell division.

Radiation

In physics, radiation describes any process in which energy emitted by one body travels through a medium or through space, ultimately to be absorbed by another body. Non-physicists often associate the word with ionizing radiation, but it can also refer to electromagnetic radiation (i.e., radio waves, infrared light, visible light, ultraviolet light, and X-rays) which can also be ionizing radiation, to acoustic radiation, or to other more obscure processes. What makes it radiation is that the energy radiates (i.e., it travels outward in straight lines in all directions) from the source.

Paleontology

Paleontology is the study of prehistoric life, including organisms' evolution and interactions with each other and their environments (their paleoecology.) As a 'historical science' it tries to explain causes rather than conduct experiments to observe effects. Paleontological observations have been documented as far back as the 5th century BC. The science became established in the 18th century as a result of Georges Cuvier's work on comparative anatomy, and developed rapidly in the 19th century.

DNA

Deoxyribonucleic acid (DNA) is a nucleic acid that contains the genetic instructions used in the development and functioning of all known living organisms and some viruses. The main role of DNA molecules is the long-term storage of information. DNA is often compared to a set of blueprints or a recipe, or a code, since it contains the instructions needed to construct other components of cells, such as proteins and RNA molecules.

Edward

The Edward mango is a monoembryonic mango cultivar grown predominantly in Florida. It is considered by many to be among the finest tasting mangoes in the world; however, its poor yields have restrained the Edward from developing into a commercially significant variety.

The Edward was first propagated in the 1920s by Edward Simmonds of the Plant Introduction Garden in Miami, Florida and is believed to be a hybrid cross of Haden and Carabao mango cultivars.

Chromosome

A Chromosome is an organized structure of DNA and protein that is found in cells. It is a single piece of coiled DNA containing many genes, regulatory elements and other nucleotide sequences. Chromosome s also contain DNA-bound proteins, which serve to package the DNA and control its functions.

Class

In biological classification, Class is

· a taxonomic rank. Other well-known ranks are life, domain, kingdom, phylum, order, family, genus, and species, with Class fitting between phylum and order. As for the other well-known ranks, there is the option of an immediately lower rank, indicated by the prefix sub-: subclass .

· a taxonomic unit, a taxon, in that rank. In that case the plural is classes

The composition of each Class is determined by a taxonomist. Often there is no exact agreement, with different taxonomists taking different positions. There are no hard rules that a taxonomist needs to follow in describing a Class, but for well-known animals there is likely to be consensus. For example, dogs are usually assigned to the Class Mammalia; in the phylum Chordata (animals with notochords); in the order Carnivora (mammals that eat meat.)

Family

In biological classification, Family is

· a taxonomic rank. Other well-known ranks are life, domain, kingdom, phylum, class, order, genus, and species, with Family fitting between order and genus. As for the other well-known ranks, there is the option of an immediately lower rank, indicated by the prefix sub-: subfamily .
· a taxonomic unit, a taxon, in that rank. In that case the plural is families

Example: 'Walnuts and Hickories belong to the Walnut Family.

What does and does not belong to each Family is determined by a taxonomist. Similarly for the question if a particular Family should be recognized at all. Often there is no exact agreement, with different taxonomists each taking a different position.

Genus

A Genus is a low-level taxonomic rank used in the classification of living and fossil organisms, and also any taxonomic unit (taxon) of that rank. The binomial name of every species is formed from a Genus name (with a capital initial), followed by the species name, both normally written in italics.

The term comes from Latin Genus 'descent, family, type, gender' , cognate with Greek: γῖνος - genos, 'race, stock, kin'.

Kingdom

In biological taxonomy, Kingdom or regnum is a taxonomic rank in either (historically) the highest rank, or (in the new three-domain system) the rank below domain. Each Kingdom is divided into smaller groups called phyla (or in some contexts these are called 'divisions'.) Currently, many textbooks from the United States use a system of six kingdoms (Animalia, Plantae, Fungi, Protista, Archaea, Bacteria) while British and Australian textbooks may describe five kingdoms (Animalia, Plantae, Fungi, Protista, and Prokaryota or Monera.)

Order

In scientific classification used in biology, the Order is

· a taxonomic rank used in the classification of organisms. Other well-known ranks are life, domain, kingdom, phylum, class, family, genus, and species, with Order fitting in between class and family. An immediately higher rank, superorder, may be added directly above Order, while suborder would be a lower rank.
· a taxonomic unit, a taxon, in that rank. In that case the plural is orders .

The Latin suffix -formes meaning 'having the form of' is used for the scientific name of orders of birds and reptiles, but not for those of mammals and invertebrates.

The Order as a distinct rank of biological classification having its own distinctive name (and not just called a higher genus (genus summum)) was first introduced by a German botanist Augustus Quirinus Rivinus in his classification of plants . Carolus Linnaeus was the first to apply it consistently to the division of all three kingdoms of Nature (minerals, plants, and animals) in his Systema Naturae (1735, 1st. Ed.).

Taxon

A Taxon (plural: taxa) is a group of (one or more) organisms, which a Taxon omist adjudges to be a unit. Usually a Taxon is given a name and a rank, although neither is a requirement. Defining what belongs or does not belong to such a Taxon omic group is done by a Taxon omist.

Technology	Technology is a broad concept that deals with an animal species' ethology or behavior of usage and of knowledge of tools and crafts, and how it affects the animal species' ability to control and adapt to its environment. Technology is a term with origins in the Greek 'technologia', 'τεχνολογΐα' -- 'techne', 'τÎχνη' and 'logia', 'λογΐα' ('saying'.) However, a strict definition is elusive; 'Technology' can refer to material objects of use to humanity, such as machines, hardware or utensils, but can also encompass broader themes, including systems, methods of organization, and techniques.
Cell	The Cell is the structural and functional unit of all known living organisms. It is the smallest unit of an organism that is classified as living, and is often called the building block of life. Some organisms, such as most bacteria, are unicellular (consist of a single Cell.)
Trait	A trait is a distinct variant of a phenotypic character of an organism that may be inherited, environmentally determined or somewhere in between. For example, eye color is a character or abstraction of an attribute, while blue, brown and hazel are traits. A trait may be any single feature or quantifiable measurement of an organism.
Archaea	The Archaea are a group of single-celled microorganisms. A single individual or species from this domain is called an archaeon (sometimes spelled 'archeon'.) They have no cell nucleus or any other organelles within their cells.
Bacteria	The Bacteria are a large group of unicellular microorganisms. Typically a few micrometres in length, Bacteria have a wide range of shapes, ranging from spheres to rods and spirals. Bacteria are ubiquitous in every habitat on Earth, growing in soil, acidic hot springs, radioactive waste, water, and deep in the Earth's crust, as well as in organic matter and the live bodies of plants and animals.

Virginia opossum	The Virginia Opossum, commonly known as the North American Opossum, is the only marsupial found in North America north of the Rio Grande River. A solitary and nocturnal animal about the size of a domestic cat, it is a successful opportunist and is found throughout Central America and North America east of the Rockies from Costa Rica to southern Ontario (it was also introduced to California in 1910, and now occupies much of the Pacific coast); it seems to be still expanding its range northward. Its ancestors evolved in South America, but were enabled to invade North America in the Great American Interchange by the formation of the Isthmus of Panama about 3 million years ago.
Anatomy	The anatomy of spiders is in some aspects similar, but also different from that of other arthropods. The following characteristics are common to all spiders: A body with two segments, eight legs, spinnerets, no chewing parts, no wings, and the presence of chelicerae, which spiders use to hold prey, and in most cases, inject venom. Spiders have non-compound eyes, with most species having eight; the spiders known as Haplogynae may have six or fewer, and certain cave-dwelling spiders may have none at all.
Capsid	A Capsid is the protein shell of a virus. It consists of several oligomeric structural subunits made of protein called protomers. The 3-dimensional morphological subunits that can be observed, which may or may not correspond to individual proteins, are called capsomeres.
Horse murders	The Horse murders scandal was a form of insurance fraud in the United States in which expensive horses, many of them show jumpers, were insured against death, accident and then killed to collect the insurance money. It is not known how many horses were murdered between the mid 1970s and the mid-1990s, when a Federal Bureau of Investigation (FBI) investigation brought the horse killings to light, but the number is thought to be well over 50, and may have been as high as 100. In addition, in 1977, the heiress Helen Brach disappeared and was presumed by law enforcement agents to have been murdered by the perpetrators of these crimes, because she threatened to report their criminal activity to authorities; continuing investigations into Brach's death began to uncover the insurance fraud in the 1990s.
Glucose	Glucose, a monosaccharide also known as grape sugar, blood sugar is a very important carbohydrate in biology. The living cell uses it as a source of energy and metabolic intermediate. Glucose is one of the main products of photosynthesis and starts cellular respiration in both prokaryotes and eukaryotes
Tobacco mosaic virus	Tobacco mosaic virus is an RNA virus that infects plants, especially tobacco and other members of the family Solanaceae. The infection causes characteristic patterns (mottling and discoloration) on the leaves (hence the name.) tobacco mosaic virus was the first virus to be discovered.
ATP synthase	An ATP synthase is a general term for an enzyme that can synthesize adenosine triphosphate (ATP) from adenosine diphosphate (ADP) and inorganic phosphate by using some form of energy. This energy is often in the form of protons moving down an electrochemical gradient, such as from the lumen into the stroma of chloroplasts or from the inter-membrane space into the matrix in mitochondria. The overall reaction sequence is: $ADP + P_i \rightarrow ATP$ These enzymes are of crucial importance in almost all organisms, because ATP is the common 'energy currency' of cells.

Bacteriophage	A Bacteriophage is any one of a number of viruses that infect bacteria. Bacteriophage s are among the most common organisms on Earth. The term is commonly used in its shortened form, phage.
Biosynthesis	Biosynthesis is a phenomenon wherein chemical compounds are produced from simpler reagents. Biosynthesis, unlike chemosynthesis, takes place within living organisms and is generally catalyzed by enzymes. The process is a vital part of metabolism.
HHV Latency Associated Transcript	HHV Latency Associated Transcript is a length of RNA which accumulates in cells hosting long-term Human Herpes Virus (HHV) infections. The LAT RNA is produced by genetic transcription from a certain region of the viral DNA. LAT regulates the viral genome and interferes with the normal activities of the infected host cell. Herpes virus may establish life-long infection during which a reservoir virus population survives in host nerve cells for long periods of time.
Lytic cycle	The Lytic cycle is one of the two cycles of viral reproduction, the other being the lysogenic cycle. These cycles should not, however, be seen as separate, but rather as somewhat interchangeable. The Lytic cycle is typically considered the main method of viral replication, since it results in the destruction of the infected cell.
Acid	An acid is traditionally considered any chemical compound that, when dissolved in water, gives a solution with a hydrogen ion activity greater than in pure water, i.e. a pH less than 7.0. That approximates the modern definition of Johannes Nicolaus Brønsted and Martin Lowry, who independently defined an acid as a compound which donates a hydrogen ion (H^+) to another compound (called a base.) Common examples include acetic acid and sulfuric acid (used in car batteries.)
Cell	The Cell is the structural and functional unit of all known living organisms. It is the smallest unit of an organism that is classified as living, and is often called the building block of life. Some organisms, such as most bacteria, are unicellular (consist of a single Cell.)
Gene	A Gene is the basic unit of heredity in a living organism. All living things depend on Gene s. Gene s hold the information to build and maintain their cells and pass Gene tic traits to offspring.
Papaya	The Papaya , is the fruit of the plant Carica Papaya, in the genus Carica. It is native to the tropics of the Americas, and was cultivated in Mexico several centuries before the emergence of the Mesoamerican classic cultures. It is sometimes called a 'big melon' or a 'paw paw' but the North American pawpaw is a different species, in the genus Asimina.
Plasmodesmata	Plasmodesmata are microscopic channels which traverse the cell walls of plant cells and some algal cells enabling transport and communication between them. Species that have Plasmodesmata include members of the Charophyceae, Charales and Coleochaetales (which are all algae), as well as all embryophytes, better known as land plants. Unlike animal cells, every plant cell is surrounded by a polysaccharide cell wall.
Reverse transcriptase	In biochemistry, a Reverse transcriptase is a DNA polymerase enzyme that transcribes single-stranded RNA into single-stranded DNA. It also helps in the formation of a double helix DNA once the RNA has been reverse transcribed into a single strand cDNA. Normal transcription involves the synthesis of RNA from DNA; hence, reverse transcription is the reverse of this.

Well studied Reverse transcriptase s include:

· HIV-1 Reverse transcriptase from human immunodeficiency virus type 1
· M-MLV Reverse transcriptase from the Moloney murine leukemia virus
· AMV Reverse transcriptase from the avian myeloblastosis virus
· Telomerase Reverse transcriptase that maintains the telomeres of eukaryotic chromosomes

The enzyme is encoded and used by reverse-transcribing viruses, which use the enzyme during the process of replication. Reverse-transcribing RNA viruses, such as retroviruses, use the enzyme to reverse-transcribe their RNA genomes into DNA, which is then integrated into the host genome and replicated along with it.

Spot

Spots are a method of smoking cannabis . In this method, small pieces of cannabis are rolled to form the 'Spot'. Generally, the tips of two knife blades are heated, the Spot is compressed between the two blades, and the subsequent smoke is inhaled.

Avian influenza

Avian influenza, sometimes avian flu, and commonly bird flu, refers to 'influenza caused by viruses adapted to birds.' Of greatest concern is highly pathogenic Avian influenza

'Bird flu' is a phrase similar to 'swine flu,' 'dog flu,' 'horse flu,' or 'human flu' in that it refers to an illness caused by any of many different strains of influenza viruses that have adapted to a specific host. All known viruses that cause influenza in birds belong to the species influenza A virus.

Viroids

Viroids are plant pathogens that consist of a short stretch (a few hundred nucleobases) of highly complementary, circular, single-stranded RNA without the protein coat that is typical for viruses. The smallest discovered is a 220 nucleobase scRNA (small cytoplasmic RNA) associated with the rice yellow mottle sobemovirus (RYMV.) In comparison, the genome of the smallest known viruses capable of causing an infection by themselves are around 2 kilobases in size.

Bacteria

The Bacteria are a large group of unicellular microorganisms. Typically a few micrometres in length, Bacteria have a wide range of shapes, ranging from spheres to rods and spirals. Bacteria are ubiquitous in every habitat on Earth, growing in soil, acidic hot springs, radioactive waste, water, and deep in the Earth's crust, as well as in organic matter and the live bodies of plants and animals.

Bacilli

Bacilli refers to a taxonomic class of bacteria. It includes two orders, Bacillales and Lactobacillales, which contain several well-known pathogens like Bacillus anthracis (the cause of anthrax.)

There are several related concepts that make use of similar words, and the ambiguity can create considerable confusion.

Binary fission

Binary fission is the form of asexual reproduction and cell division used by all prokaryotic and some eukaryotic organisms. This process results in the reproduction of a living prokaryotic cell by division into two parts which each have the potential to grow to the size of the original cell.

Mitosis and cytokinesis are not the same as Binary fission.

Nucleoid

The Nucleoid is an irregularly-shaped region within the cell of prokaryotes which has nuclear material without a nuclear membrane and where the genetic material is localized. The genome of prokaryotic organisms generally is a circular, double-stranded piece of DNA, of which multiple copies may exist at any time. The length of a genome widely varies, but generally is at least a few million base pairs.

Peptidoglycan

Peptidoglycan is a polymer consisting of sugars and amino acids that forms a mesh-like layer outside the plasma membrane of bacteria, forming the cell wall. The sugar component consists of alternating residues of β- linked N-acetylglucosamine and N-acetylmuramic acid residues. Attached to the N-acetylmuramic acid is a peptide chain of three to five amino acids.

Plasmid

A Plasmid is an extra-chromosomal DNA molecule separate from the chromosomal DNA which is capable of replicating independently of the chromosomal DNA. In many cases, it is circular and double-stranded. Plasmid s usually occur naturally in bacteria, but are sometimes found in eukaryotic organisms (e.g., the 2-micrometre-ring in Saccharomyces cerevisiae.)

Vibrio

Vibrio is a genus of Gram-negative bacteria possessing a curved rod shape. Typically found in saltwater, Vibrio are facultative anaerobes that test positive for oxidase and do not form spores. All members of the genus are motile and have polar flagella with sheaths.

CVS

CVS is a terpene cyclase enzyme responsible for the biosynthesis of valencene, a sesquiterpene, using farnesyl pyrophosphate as its substrate. The first CVS enzyme was isolated using orange cDNA. .

Anthrax

Anthrax is an acute disease caused by Bacillus anthracis. It affects both humans and animals and most forms of the disease are highly lethal. There are effective vaccines against Anthrax, and some forms of the disease respond well to antibiotic treatment.

Autotroph

An autotroph is an organism that produces complex organic compounds from simple inorganic molecules using energy from light (by photosynthesis) or inorganic chemical reactions.

autotroph s are the producers in a food chain, such as plants on land or algae in water. Bacteria which derive energy from oxidizing inorganic compounds (such as hydrogen sulfide, ammonium and ferrous iron) are chemo autotroph s, and include the lithotrophs.

Chromosome

A Chromosome is an organized structure of DNA and protein that is found in cells. It is a single piece of coiled DNA containing many genes, regulatory elements and other nucleotide sequences. Chromosome s also contain DNA-bound proteins, which serve to package the DNA and control its functions.

Endospore

An Endospore is a dormant, tough, and non-reproductive structure produced by Gram-positive bacteria from the Firmicute phylum which forms when a bacterium produces a thick internal wall that encloses its DNA and part of its cytoplasm. Examples include Bacillus and Clostridium.

The primary function of most Endospore s is to ensure the survival of a bacterium through periods of environmental stress.

Heterochromatin

Heterochromatin is a tightly packed form of DNA. Its major characteristic is that transcription is limited. As such, it is a means to control gene expression, through regulation of the transcription initiation.

Chromatin is found in two varieties: euchromatin and Heterochromatin.

Heterotroph

A Heterotroph is an organism that uses organic substrates to get its chemical energy for its life cycle. This contrasts with autotrophs such as plants which are able to directly use sources of energy such as light to produce organic substrates from inorganic carbon dioxide. The Cyanobacteria Synechocystis sp.

Transduction

Transduction is the process by which DNA is transferred from one bacterium to another by a virus. It also refers to the process whereby foreign DNA is introduced into another cell via a viral vector. This is a common tool used by molecular biologists to stably introduce a foreign gene into a host cell's genome.

Nereis

Nereis is a genus of polychaete worms in the family Nereidae. It comprises many species, most of which are marine, including the sandworm (Nereis virens) and the common clam worm (Nereis succinea.) Nereis possess setae and parapodia for locomotion.

Bioremediation

Bioremediation can be defined as any process that uses microorganisms, fungi, green plants or their enzymes to return the natural environment altered by contaminants to its original condition. Bioremediation may be employed to attack specific soil contaminants, such as degradation of chlorinated hydrocarbons by bacteria. An example of a more general approach is the cleanup of oil spills by the addition of nitrate and/or sulfate fertilisers to facilitate the decomposition of crude oil by indigenous or exogenous bacteria.

Decomposers

Decomposers are organisms that consume dead or decaying organisms, and, in doing so, carry out the natural process of decomposition. Like herbivores and predators, Decomposers are heterotrophic, meaning that they use organic substrates to get their energy, carbon and nutrients for growth and development. Decomposers use deceased organisms and non-living organic compounds as their food source.

Decomposition

Decomposition refers to the process by which tissues of a dead organism breaks down into simpler forms of matter. Such a breakdown of dead organisms is essential for new growth and development of living organisms because it recycles the finite matter that occupies physical space in the biome. Bodies of living organisms begin to decompose shortly after death.

Nitrogen

Nitrogen is a chemical element that has the symbol N and atomic number 7 and atomic mass 14.00674 u. Elemental Nitrogen is a colorless, odorless, tasteless and mostly inert diatomic gas at standard conditions, constituting 78% by volume of Earth's atmosphere.

Many industrially important compounds, such as ammonia, nitric acid, organic nitrates , and cyanides, contain Nitrogen.

Nitrogen fixation

Nitrogen fixation is the process by which nitrogen is taken from its relatively inert molecular form (N_2) in the atmosphere and converted into nitrogen compounds (such as ammonia, nitrate and nitrogen dioxide.) This is an essential process for life because fixed nitrogen is needed to make nucleotides which are needed to make DNA and also to make amino acids which in turn are needed to produce proteins.

Nitrogen fixation is performed naturally by a number of different prokaryotes, including bacteria, actinobacteria, and certain types of anaerobic bacteria.

Clostridium	Clostridium is a genus of Gram-positive bacteria, belonging to the Firmicutes. They are obligate anaerobes capable of producing endospores. Individual cells are rod-shaped, which gives them their name, from the Greek kloster or spindle.
Salmonella	Salmonella is a genus of rod-shaped, Gram-negative, non-spore forming, predominantly motile enterobacteria with diameters around 0.7 to 1.5 Âµm, lengths from 2 to 5 Âµm, and flagella which project in all directions (i.e. peritrichous.) They are chemoorganotrophs, obtaining their energy from oxidation and reduction reactions using organic sources and are facultative anaerobes; most species produce hydrogen sulfide, which can readily be detected by growing them on media containing ferrous sulfate, such as TSI. Most isolates exist in two phases; phase I is the motile phase and phase II the non-motile phase. Cultures that are non-motile upon primary culture may be swithched to the motile phase using a Craigie tube.
Shigella	Shigella is a genus of Gram-negative, non-spore forming rod-shaped bacteria closely related to Escherichia coli and Salmonella. The causative agent of human shigellosis, Shigella cause disease in primates, but not in other mammals. It is only naturally found in humans and apes.
Shigella dysenteriae	Shigella dysenteriae is a species of the rod-shaped bacterial genus Shigella. Shigella can cause shigellosis (bacillary dysentery.) Shigellae are Gram-negative, non-spore-forming, facultatively anaerobic, non-motile bacteria.
Fermentation	Fermentation is the process of deriving energy from the oxidation of organic compounds, such as carbohydrates, using an endogenous electron acceptor, which is usually an organic compound. This is in contrast to cellular respiration, where electrons are donated to an exogenous electron acceptor, such as oxygen, via an electron transport chain. Fermentation does not necessarily have to be carried out in an anaerobic environment.
Human	A Human is a member of a species of bipedal primates in the family Hominidae . DNA and fossil evidence indicates that modern Human s originated in east Africa about 200,000 years ago. When compared to other animals and primates, Human s have a highly developed brain, capable of abstract reasoning, language, introspection and problem solving.
Tetanus	Tetanus is a medical condition characterized by a prolonged contraction of skeletal muscle fibers. The primary symptoms are caused by tetanospasmin, a neurotoxin produced by the Gram-positive, obligate anaerobic bacterium Clostridium tetani. Infection generally occurs through wound contamination and often involves a cut or deep puncture wound.
Operon	An Operon is a functioning unit of key nucleotide sequences of DNA including an operator, a common promoter, and one or more structural genes, which is controlled as a unit to produce messenger RNA (mRNA), in the process of transcription by an RNA polymerase. A typical Operon. The term "Operon" was first proposed in a short paper in the Proceedings of the French Academy of Science in 1960. From this paper, the so-called general theory of the Operon was developed.
Pathogen	A pathogen 'I give birth to'), infectious agent is a biological agent that causes disease or illness to its host. There are several substrates and pathways whereby pathogen s can invade a host; the principal pathways have different episodic time frames, but soil contamination has the longest or most persistent potential for harboring a pathogen

The body contains many natural defenses against some of the common pathogen s (such as Pneumocystis) in the form of the human immune system and by some 'helpful' bacteria present in the human body's normal flora.

Toxin

A Toxin is a poisonous substance produced by living cells or organisms.

For a toxic substance not produced by living organisms, 'toxicant' is the more appropriate term, and 'toxics' is an acceptable plural.

Toxin s can be small molecules, peptides, or proteins that are capable of causing disease on contact with or absorption by body tissues interacting with biological macromolecules such as enzymes or cellular receptors.

Typhoid fever

Typhoid fever Salmonella typhi or commonly just typhoid, is an illness caused by the bacterium Salmonella enterica serovar typhi. Common worldwide, it is transmitted by the ingestion of food or water contaminated with feces from an infected person. The bacteria then perforate through the intestinal wall and are phagocytosed by macrophages.

APC

APC (adenomatosis polyposis coli) is a human gene that is classified as a tumor suppressor gene. Tumor suppressor genes prevent the uncontrolled growth of cells that may result in cancerous tumors. The protein made by the APC gene plays a critical role in several cellular processes that determine whether a cell may develop into a tumor.

Archaea

The Archaea are a group of single-celled microorganisms. A single individual or species from this domain is called an archaeon (sometimes spelled 'archeon'.) They have no cell nucleus or any other organelles within their cells.

Cloning

Cloning in biology is the process of producing populations of genetically-identical individuals that occurs in nature when organisms such as bacteria, insects or plants reproduce asexually. Cloning in biotechnology refers to processes used to create copies of DNA fragments (molecular Cloning), cells (cell Cloning), or organisms. More generally, the term refers to the production of multiple copies of a product such as digital media or software.

Halophiles

Halophiles are extremophile organisms that thrive in environments with very high concentrations of salt. The name comes from Greek for 'salt-loving'. While the term is perhaps most often applied to some Halophiles classified into the Archaea domain, there are also bacterial Halophiles and some eukaryota, such as the alga Dunaliella salina.

Methanogens

Methanogens are archaea that produce methane as a metabolic byproduct in anoxic conditions. They are common in wetlands, where they are responsible for marsh gas, and in the guts of animals such as ruminants and humans, where they are responsible for the methane content of flatulence. In marine sediments biomethanation is generally confined to where sulfates are depleted, below the top layers.

Thermoacidophile

A Thermoacidophile (combination of thermophile and acidophile) is an extreme archeon which thrives in acidous, sulfur rich, high temperature environments.

Thermoacidophile s prefer temperatures of 70 - 80 °C and pH between 2 and 3. They live mostly in hot springs and/or within deep ocean vent communities.

Alagille syndrome

Alagille syndrome is a genetic disorder that affects the liver, heart, and other systems of the body. Problems associated with the disorder generally become evident in infancy or early childhood. The disorder is inherited in an autosomal dominant pattern, and the estimated prevalence of Alagille syndrome is 1 in every 100,000 live births.

Escherichia

Escherichia is a genus of Gram-negative, non-spore forming, facultatively anaerobic, rod-shaped bacteria from the family Enterobacteriaceae. Inhabitants of the gastrointestinal tracts of warm-blooded animals, Escherichia species provide a portion of the microbially-derived vitamin K for their host.

While many Escherichia are harmless commensals, particular strains of some species are human pathogens, and are known as the most common cause of urinary tract infections, significant sources of gastrointestinal disease, ranging from simple diarrhea to dysentery-like conditions, as well as a wide-range of other pathogenic states.

Escherichia coli

Escherichia coli , is a Gram negative bacterium that is commonly found in the lower intestine of warm-blooded organisms . Most E. coli strains are harmless, but some, such as serotype O157:H7, can cause serious food poisoning in humans, and are occasionally responsible for costly product recalls. The harmless strains are part of the normal flora of the gut, and can benefit their hosts by producing vitamin K_2, or by preventing the establishment of pathogenic bacteria within the intestine.

Chlamydomonas

Chlamydomonas is a genus of green alga. They are unicellular flagellates. Chlamydomonas is used as a model organism for molecular biology, especially studies of flagellar motility and chloroplast dynamics, biogenesis, and genetics.

Spirogyra

Spirogyra is a genus of filamentous green algae of the order Zygnematales and there are more than 400 species of Spirogyra in the world. Spirogyra measures approximately 10 to 100µm in width and may stretch centimeters long.

Spirogyra is unbranched with cylindrical cells connected end to end in long green filaments.

Volvox

Volvox is one of the best-known chlorophytes and is the most developed in a series of genera that form spherical colonies. Each mature Volvox colony is composed of numerous flagellate cells similar to Chlamydomonas, up to 50,000 in total, and embedded in the surface of a hollow sphere or coenobium containing an extracellular matrix made of a gelatinous glycoprotein. The cells swim in a coordinated fashion, with a distinct anterior and posterior poles.

Green algae

The Green algae are the large group of algae from which the embryophytes (higher plants) emerged. As such, they form a paraphyletic group, although the group including both Green algae and embryophytes is monophyletic (and often just known as kingdom Plantae.) The Green algae include unicellular and colonial flagellates, usually but not always with two flagella per cell, as well as various colonial, coccoid, and filamentous forms.

Punnett square

The Punnett square is a diagram that is used to predict the outcome of a particular cross or breeding experiment. It is named after Reginald C. Punnett, who devised the approach, and is used by biologists to determine the probability of an offspring having a particular genotype. The Punnett square is a summary of every possible combination of one maternal allele with one paternal allele for each gene being studied in the cross.

Amoeboid

Amoeboid s are unicellular life-forms characterized by their irregularity of shape.

Amoeboid and 'amoeba' are sometimes used interchangeably in less formal contexts, especially in the context of characterizing an organism by the method of locomotion.

Amoeboid s are unicellular life-forms characterized by their similarity to amoebae.

Hormone	Hormone s are chemicals released by cells that affect cells in other parts of the body. Only a small amount of Hormone is required to alter cell metabolism. It is essentially a chemical messenger that transports a signal from one cell to another.
Pseudopods	Pseudopods or pseudopodia (singular: pseudopodium) are temporary projections of eukaryotic cells. Cells having this faculty are generally referred to as amoeboids. Pseudopodia extend and contract by the reversible assembly of actin subunits into microfilaments.
Blight	Blight refers to a specific symptom affecting plants in response to infection by a pathogenic organism. It is simply a rapid and complete chlorosis, browning, then death of plant tissues such as leaves, branches, twigs, or floral organs. Accordingly, many diseases that primarily exhibit this symptom are called Blight s.
Potassium	Potassium is a chemical element. It has the symbol K , atomic number 19, and atomic mass 39.0983. Potassium was first isolated from potash.

Alagille syndrome

Alagille syndrome is a genetic disorder that affects the liver, heart, and other systems of the body. Problems associated with the disorder generally become evident in infancy or early childhood. The disorder is inherited in an autosomal dominant pattern, and the estimated prevalence of Alagille syndrome is 1 in every 100,000 live births.

Chara

Chara species are multicellular and superficially resemble land plants because of stem-like and leaf-like structures. The branching system is complex with branches derived from apical cells which cut off segments at the base to form nodal and internodal cells alternately. They are typically anchored to the littoral substrate by means of branching underground rhizoids.

Coleochaete

Coleochaete is a genus of filamentous green algae in the family Coleochaetaceae. They are haploid, reproduce asexually, live in freshwater environments worldwide
· Coleochaete irregularis (TSN 9360)
· Coleochaete nitellarum (TSN 9365)
· Coleochaete orbicularis (TSN 9363)
· Coleochaete pulvinata (TSN 9361)
· Coleochaete sampsonii (TSN 9366)
· Coleochaete scutata (TSN 9364)
· Coleochaete soluta (TSN 9362)

.

Plants

Plants are living organisms belonging to the kingdom Plantae. They include familiar organisms such as trees, herbs, bushes, grasses, vines, ferns, mosses, and green algae. About 350,000 species of Plants, defined as seed Plants, bryophytes, ferns and fern allies, are estimated to exist currently.

Seed

A Seed, referred to as a kernel in some plants, is a small embryonic plant enclosed in a covering called the Seed coat, usually with some stored food. It is the product of the ripened ovule of gymnosperm and angiosperm plants which occurs after fertilization and some growth within the mother plant. The formation of the Seed completes the process of reproduction in Seed plants (started with the development of flowers and pollination), with the embryo developed from the zygote and the Seed coat from the integuments of the ovule.

Flower

A Flower sometimes known as a bloom or blossom, is the reproductive structure found in Flower ing plants The biological function of a Flower is to mediate the union of male sperm with female ovum in order to produce seeds. The process begins with pollination, is followed by fertilization, leading to the formation and dispersal of the seeds.

DNA

Deoxyribonucleic acid (DNA) is a nucleic acid that contains the genetic instructions used in the development and functioning of all known living organisms and some viruses. The main role of DNA molecules is the long-term storage of information. DNA is often compared to a set of blueprints or a recipe, or a code, since it contains the instructions needed to construct other components of cells, such as proteins and RNA molecules.

Spirogyra

Spirogyra is a genus of filamentous green algae of the order Zygnematales and there are more than 400 species of Spirogyra in the world. Spirogyra measures approximately 10 to 100μm in width and may stretch centimeters long.

Spirogyra is unbranched with cylindrical cells connected end to end in long green filaments.

Alternation of phases	The Alternation of phases describes the life cycle of plants, fungi and protists. A multicellular diploid phase alternates with a multicellular haploid phase. The term can be confusing for people familiar only with the life cycle of a typical animal.
Horse murders	The Horse murders scandal was a form of insurance fraud in the United States in which expensive horses, many of them show jumpers, were insured against death, accident and then killed to collect the insurance money. It is not known how many horses were murdered between the mid 1970s and the mid-1990s, when a Federal Bureau of Investigation (FBI) investigation brought the horse killings to light, but the number is thought to be well over 50, and may have been as high as 100. In addition, in 1977, the heiress Helen Brach disappeared and was presumed by law enforcement agents to have been murdered by the perpetrators of these crimes, because she threatened to report their criminal activity to authorities; continuing investigations into Brach's death began to uncover the insurance fraud in the 1990s.
Gametophyte	In plants and algae that undergo alternation of generations, a Gametophyte is the multicellular structure, or phase, that is haploid, containing a single set of chromosomes: The Gametophyte produces male or female gametes (or both), by a process of cell division called mitosis. The fusion of male and female gametes produces a diploid zygote, which develops by repeated mitotic cell divisions into a multicellular sporophyte. Because the sporophyte is the product of the fusion of two haploid gametes, its cells are diploid, containing two sets of chromosomes.
Gymnosperm	Gymnosperm (Gymnosperm ae) is a group of spermatophyte seed-bearing plants with ovules on scales, which are usually arranged in cone-like structures. The other major group of seed-bearing plants, the angiosperms , have ovules enclosed in a carpel, a sporophyll with fused margins. A carpel consists of a stigma, style and ovary.
Size	Size has been one of the most interesting aspects of cephalopod science to the general public Extinct taxa are also included.
Spore	In biology, a Spore is a reproductive structure that is adapted for dispersal and surviving for extended periods of time in unfavorable conditions. Spore s form part of the life cycles of many bacteria, plants, algae, fungi and some protozoans. A chief difference between Spore s and seeds as dispersal units is that Spore s have very little stored food resources compared with seeds.
Sporophyte	All land plants, and some algae, have life cycles in which a haploid gametophyte generation alternates with a diploid Sporophyte, the generation of a plant or alga that has a double set of chromosomes. A multicellular Sporophyte generation or phase is present in the life cycle of all land plants and in some green algae. For common flowering plants (Angiosperms), the Sporophyte generation comprises almost their whole life cycle (i.e. whole green plant, roots etc), except phases of small reproductive structures (pollen and ovule.)
Technology	Technology is a broad concept that deals with an animal species' ethology or behavior of usage and of knowledge of tools and crafts, and how it affects the animal species' ability to control and adapt to its environment. Technology is a term with origins in the Greek 'technologia', 'τεχνολογΓ¯α' -- 'techne', 'τÎχνη' and 'logia', 'λογΓ¯α' ('saying'.) However, a strict definition is elusive; 'Technology' can refer to material objects of use to humanity, such as machines, hardware or utensils, but can also encompass broader themes, including systems, methods of organization, and techniques.

Dehiscence

Dehiscence is the spontaneous opening at maturity of a plant structure, such as a fruit, anther to release its contents. This is the final function of the anther that causes the release of pollen grains. The anther wall breaks at a specific site that runs the length of the anther.

Clostridium

Clostridium is a genus of Gram-positive bacteria, belonging to the Firmicutes. They are obligate anaerobes capable of producing endospores. Individual cells are rod-shaped, which gives them their name, from the Greek kloster or spindle.

Chromosome

A Chromosome is an organized structure of DNA and protein that is found in cells. It is a single piece of coiled DNA containing many genes, regulatory elements and other nucleotide sequences. Chromosome s also contain DNA-bound proteins, which serve to package the DNA and control its functions.

Club

In zoology, a Club is a bony mass at the end of the tail of some dinosaurs and of some mammals, most notably the ankylosaurids and the glyptodonts. It is thought that this was a form of defensive armour or weapon that was used to defend against predators, much in the same way as a thagomizer, possessed by stegosaurids, though at least in glyptodonts it is hypothesized it was used in fighting for mating rights. Among dinosaurs, the Club was present mainly in ankylosaurids, although the sauropod Shunosaurus also possessed a tail Club.

Fern

A Fern is any one of a group of about 20,000 species of plants classified in the phylum or division Pteridophyta, also known as Filicophyta. The group is also referred to as Polypodiophyta, or Polypodiopsida when treated as a subdivision of tracheophyta (vascular plants.) The term 'pteridophyte' has traditionally been used to describe all seedless vascular plants, making it synonymous with ' Fern s and Fern allies'.

Frond

A Frond is a large leaf with many divisions to it, and the term is typically used for the leaves of palms, ferns or cycads.

A Frond is the leaf- like structure of a fern or alga. The term is colloquially applied to the leaves of palms, cycads, and plants with pinnately compound leaves.

Light-dependent reactions

The Light-dependent reactions are the first stage of photosynthesis. In this process light energy is converted into chemical energy, in the form of the energy-carriers ATP and NADPH. In the light-independent reactions, the formed NADPH and ATP drive the reduction of CO_2 to more useful organic compounds, such as glucose.

The Light-dependent reactions take place on the thylakoid membrane inside a chloroplast.

Microphyll

Traditionally, a Microphyll is 'an appendage supplied by a single, unbranched vein'. Despite their name, Microphyll s are not always microscopic; those of Isoetes (quillworts) reach centimetres in length, and the extinct Lepidodendron bore Microphyll s over a metre long. In the classical concept of a Microphyll this vein emerges from the protostele, without leaving a gap.

Phloem

In vascular plants, Phloem is the living tissue that carries organic nutrients (known as photosynthate), particularly sucrose, a sugar, to all parts of the plant where needed. In trees, the Phloem is the innermost layer of the bark, hence the name, derived from the Greek word φλΪŒος 'bark'. The Phloem is mainly concerned with the transport of soluble organic material made during photosynthesis.

Vascular plants	Vascular plants are those plants that have lignified tissues for conducting water, minerals, and photosynthetic products through the plant. Vascular plants include the ferns, clubmosses, flowering plants, conifers and other gymnosperms. Scientific names for the group include Tracheophyta and Tracheobionta, but neither name is very widely used.
Vascular tissue	Vascular tissue is a complex conducting tissue, formed of more than one cell type, found in vascular plants. The primary components of Vascular tissue are the xylem and phloem. These two tissues transport fluid and nutrients internally.
Xylem	In vascular plants, xylem is one of the two types of transport tissue, phloem being the other. The word 'xylem' is derived from classical Greek ξυλον , 'wood', and indeed the best known xylem tissue is wood, though it is found throughout the plant. Its basic function is to transport water.
Pollen	Pollen is a fine to coarse powder consisting of microgametophytes (Pollen grains), which produce the male gametes (sperm cells) of seed plants. A hard coat covering the Pollen grain protects the sperm cells during the process of their movement between the stamens of the flower to the pistil of the next flower. Individual Pollen grains are small enough to require magnification to see detail.
Pollination	Pollination is the process by which pollen is transferred in plants, thereby enabling fertilisation and sexual reproduction. Pollen grains, which contain the male gametes (sperm) to where the female gamete(s) are contained within the carpel; in gymnosperms the pollen is directly applied to the ovule itself. The receptive part of the carpel is called a stigma in the flowers of angiosperms.
Cone	A cone is an organ on plants in the division Pinophyta (conifers) that contains the reproductive structures. The familiar woody cone is the female cone, which produces seeds. The male cones, which produce pollen, are usually herbaceous and much less conspicuous even at full maturity.
Calvin cycle	The Calvin cycle is a series of biochemical reactions that take place in the stroma of chloroplasts in photosynthetic organisms. It was discovered by Melvin Calvin, James Bassham and Andrew Benson at the University of California, Berkeley . It is one of the light-independent reactions or dark reactions.
Anatomy	The anatomy of spiders is in some aspects similar, but also different from that of other arthropods. The following characteristics are common to all spiders: A body with two segments, eight legs, spinnerets, no chewing parts, no wings, and the presence of chelicerae, which spiders use to hold prey, and in most cases, inject venom. Spiders have non-compound eyes, with most species having eight; the spiders known as Haplogynae may have six or fewer, and certain cave-dwelling spiders may have none at all.
Calyx	Calyx is a term used in animal anatomy for some cuplike areas or structures. The spicules containing the basal portion of the upper tentacular part of the polyp of some soft corals (also called calice.) A body part of the Entoprocta from which tentacles arise and the mouth and anus are located.
Carpel	Carpel s are the building blocks of the gynoecium. If a gynoecium has a single carpel it is called monocarpous. If a gynoecium has multiple, distinct (free, unfused) carpel s, it is apocarpous.

Flowering

A flower, sometimes known as a bloom or blossom, is the reproductive structure found in flowering plants The biological function of a flower is to mediate the union of male sperm with female ovum in order to produce seeds. The process begins with pollination, is followed by fertilization, leading to the formation and dispersal of the seeds.

Flowering plants

The Flowering plants or angiosperms (Angiospermae or Magnoliophyta) are the most widespread group of land plants. The Flowering plants and the gymnosperms are the only extant groups of seed plants. The Flowering plants are distinguished from other seed plants by a series of apomorphies, or derived characteristics.

Ovary

In the flowering plants, an Ovary is a part of the female reproductive organ of the flower or gynoecium. Specifically, it is the part of the carpel which holds the ovule(s) and is located above or below or at the point of connection with the base of the petals and sepals. In this picture of a zucchini the petals and sepals are above the Ovary and such a flower is said to have an inferior Ovary; also referred to as epigynous.

Sepal

A Sepal is a part of the flower of angiosperms . Sepal s in most flowers are green and lie under the more conspicuous petals. As a collective unit the Sepal s form a calyx, whereas the collection of petals is called the corolla.

Cotyledon

A Cotyledon is a significant part of the embryo within the seed of a plant. Upon germination, the Cotyledon may become the embryonic first leaves of a seedling. The number of Cotyledon s present is one characteristic used by botanists to classify the flowering plants .

Double fertilization

The Parts of a Flower. double fertilization.

double fertilization is a complex fertilization mechanism that has evolved in flowering plants, known as angiosperms. This process involves the joining of a female gametophyte (embryo sac) with two male gametes (sperm.) It begins when a pollen grain adheres to the stigma of the carpel, the female reproductive structure of a flower.

Endosperm

Endosperm is the tissue produced in the seeds of most flowering plants around the time of fertilization. It surrounds the embryo and provides nutrition in the form of starch, though it can also contain oils and protein. This makes Endosperm an important source of nutrition in human diet.

Fruit

The term fruit has different meanings dependent on context, and the term is not synonymous in food preparation and biology. fruit s are the means by which flowering plants disseminate seeds, and the presence of seeds indicates that a structure is most likely a fruit though not all seeds come from fruit s.

No single terminology really fits the enormous variety that is found among plant fruit s.

Pollen tube

The Pollen tube of most seed plants acts as a conduit to transport sperm cells from the pollen grain, either from the stigma (in flowering plants or angiosperms) to the ovules at the base of the pistil and does not convey sperm cells to the egg. Like ferns, other basal land plants, and many algae, these gymnosperms have flagellate sperm, which swim through a watery fluid to fertilize the egg cells.

Adaptation

In ocular physiology, Adaptation is the ability of the eye to adjust to various levels of darkness and light.

The human eye can function from very dark to very bright levels of light -- its sensing capabilities reach across nine orders of magnitude. This means that the brightest and the darkest light signal that the eye can sense are a factor of roughly one thousand million apart.

Cotton

Cotton is a soft, staple fiber that grows in a form known as a boll around the seeds of the Cotton plant, a shrub native to tropical and subtropical regions around the world, including the Americas, India and Africa. The fiber most often is spun into yarn or thread and used to make a soft, breathable textile, which is the most widely used natural-fiber cloth in clothing today. The English name derives from the Arabic qutn Ù‚Ø·Ù† , which began to be used circa 1400.

Pollinator

A Pollinator is the biotic agent (vector) that moves pollen from the male anthers of a flower to the female stigma of a flower to accomplish fertilization or syngamy of the female gamete in the ovule of the flower by the male gamete from the pollen grain. Though the terms are sometimes confused, a Pollinator is different from a pollenizer, which is a plant that is a source of pollen for the pollination process.

Plants fall into pollination syndromes that reflect the type of Pollinator being attracted.

Fungi

A fungus is a eukaryotic organism that is a member of the kingdom fungi . The fungi are a monophyletic group, also called the Eumycota , that is phylogenetically distinct from the structurally similar slime molds (myxomycetes) and water molds (oomycetes.) The fungi are heterotrophic organisms possessing a chitinous cell wall, with most species growing as multicellular filaments called hyphae forming a mycelium; some species also grow as single cells.

Mycelium

Mycelium is the vegetative part of a fungus, consisting of a mass of branching, thread-like hyphae. The mass of hyphae is sometimes called shiro, especially within the fairy ring fungi. Fungal colonies composed of mycelia are found in soil and on or in many other substrates.

Rhizopus stolonifer

Rhizopus stolonifer is a widely distributed Mucoralean mold. Commonly found on bread surfaces, it takes food and nutrients from the bread and causes damage to the surface where it lives.

Asexual spores are formed within sporangia, which break to release the spores mature.

Oyster

The common name Oyster is used for a number of different groups of bivalve mollusks, most of which live in marine habitats or brackish water. The shell consists of two usually highly calcified valves which surround a soft body. Gills filter plankton from the water, and strong adductor muscles are used to hold the shell closed.

Decomposers

Decomposers are organisms that consume dead or decaying organisms, and, in doing so, carry out the natural process of decomposition. Like herbivores and predators, Decomposers are heterotrophic, meaning that they use organic substrates to get their energy, carbon and nutrients for growth and development. Decomposers use deceased organisms and non-living organic compounds as their food source.

Y chromosome

The Y chromosome is the sex-determining chromosome in most mammals, including humans. In mammals, it contains the gene SRY, which triggers testis development, thus determining sex. The human Y chromosome is composed of about 60 million base pairs.

Fermentation

Fermentation is the process of deriving energy from the oxidation of organic compounds, such as carbohydrates, using an endogenous electron acceptor, which is usually an organic compound. This is in contrast to cellular respiration, where electrons are donated to an exogenous electron acceptor, such as oxygen, via an electron transport chain. Fermentation does not necessarily have to be carried out in an anaerobic environment.

Crassulacean acid metabolism	Crassulacean acid metabolism is an elaborate carbon fixation pathway in some plants. These plants fix carbon dioxide during the night, storing it as the four carbon acid malate. The CO_2 is released during the day, where it is concentrated around the enzyme RuBisCO, increasing the efficiency of photosynthesis.
Candida	Candida is a genus of yeasts. Many species of this genus are endosymbionts of animal hosts including humans. While usually living as commensals, some Candida species have the potential to cause disease.
Chytridiomycosis	Chytridiomycosis is an infectious disease of amphibians, caused by the chytrid Batrachochytrium dendrobatidis, a non-hyphal zoosporic fungus. Chytridiomycosis has been linked to dramatic population declines or even extinctions of amphibian species in western North America, Central America, South America, eastern Australia, and Dominica and Montserrat in the Caribbean. The fungus is capable of causing sporadic deaths in some amphibian populations and 100% mortality in others.
Gene	A Gene is the basic unit of heredity in a living organism. All living things depend on Gene s. Gene s hold the information to build and maintain their cells and pass Gene tic traits to offspring.
Trehalose	Trehalose is a natural alpha-linked disaccharide formed by an α, α-1, 1-glucoside bond between two α-glucose units. In 1832 Wiggers discovered Trehalose in an ergot of rye and in 1859 Berthelot isolated it from trehala manna, a substance made by weevils, and named it Trehalose. It can be synthesised by fungi, plants, and invertebrate animals.

Animals	Animals are a major group of mostly multicellular, eukaryotic organisms of the kingdom Animalia or Metazoa. Their body plan eventually becomes fixed as they develop, although some undergo a process of metamorphosis later on in their life. Most Animals are motile, meaning they can move spontaneously and independently.
Escherichia	Escherichia is a genus of Gram-negative, non-spore forming, facultatively anaerobic, rod-shaped bacteria from the family Enterobacteriaceae. Inhabitants of the gastrointestinal tracts of warm-blooded animals, Escherichia species provide a portion of the microbially-derived vitamin K for their host. While many Escherichia are harmless commensals, particular strains of some species are human pathogens, and are known as the most common cause of urinary tract infections, significant sources of gastrointestinal disease, ranging from simple diarrhea to dysentery-like conditions, as well as a wide-range of other pathogenic states.
Escherichia coli	Escherichia coli , is a Gram negative bacterium that is commonly found in the lower intestine of warm-blooded organisms . Most E. coli strains are harmless, but some, such as serotype O157:H7, can cause serious food poisoning in humans, and are occasionally responsible for costly product recalls. The harmless strains are part of the normal flora of the gut, and can benefit their hosts by producing vitamin K_2, or by preventing the establishment of pathogenic bacteria within the intestine.
Punnett square	The Punnett square is a diagram that is used to predict the outcome of a particular cross or breeding experiment. It is named after Reginald C. Punnett, who devised the approach, and is used by biologists to determine the probability of an offspring having a particular genotype. The Punnett square is a summary of every possible combination of one maternal allele with one paternal allele for each gene being studied in the cross.
Metamorphosis	Metamorphosis is a biological process by which an animal physically develops after birth or hatching, involving a conspicuous and relatively abrupt change in the animal's form or body structure through cell growth and differentiation. Some insects, amphibians, mollusks, crustaceans, Cnidarians, echinoderms and tunicates undergo Metamorphosis, which is usually (but not always) accompanied by a change of habitat or behavior. Scientific usage of the term is exclusive, and is not applied to general aspects of cell growth, including rapid growth spurts.
Acetabularia	Acetabularia is a genus of green algae, specifically of the Polyphysaceae family, Typically found in subtropical waters, Acetabularia is a single-cell organism, but gigantic in size and complex in form, making it an excellent model organism for studying cell biology. In form, the mature Acetabularia resembles the round leaves of a nasturtium, being 0.5 to 10 cm tall and having three anatomical parts: a bottom rhizoid that resembles a set of short roots; a long stalk in the middle; and a top umbrella of branches that may fuse into a cap. The single nucleus of Acetabularia is located in the rhizoid, and allows the cell to regenerate completely if its cap is removed.
Acetylcholine	The chemical compound Acetylcholine is a neurotransmitter in both the peripheral nervous system (PNS) and central nervous system (CNS) in many organisms including humans. Acetylcholine is one of many neurotransmitters in the autonomic nervous system (ANS) and the only neurotransmitter used in the motor division of the somatic nervous system. (Sensory neurons use glutamate and various peptides at their synapses.)
Coelom	The Coelom is a fluid filled cavity formed within the mesoderm. Coelom s developed in triploblasts but were subsequently lost in several lineages. Loss of Coelom is correlated with reduction in body size.

Decomposers	Decomposers are organisms that consume dead or decaying organisms, and, in doing so, carry out the natural process of decomposition. Like herbivores and predators, Decomposers are heterotrophic, meaning that they use organic substrates to get their energy, carbon and nutrients for growth and development. Decomposers use deceased organisms and non-living organic compounds as their food source.
Deuterostomes	Deuterostomes are a superphylum of animals. They are a subtaxon of the Bilateria branch of the subregnum Eumetazoa, and are opposed to the protostomes. Deuterostomes are distinguished by their embryonic development; in Deuterostomes, the first opening becomes the anus, while in protostomes it becomes the mouth.
Protostomes	Protostomia are a clade of animals. Together with the deuterostomes and a few smaller phyla, they make up the Bilateria, mostly comprising animals with bilateral symmetry and three germ layers. The major distinctions between deuterostomes and Protostomes are found in embryonic development.
Clostridium	Clostridium is a genus of Gram-positive bacteria, belonging to the Firmicutes. They are obligate anaerobes capable of producing endospores. Individual cells are rod-shaped, which gives them their name, from the Greek kloster or spindle.
Spirogyra	Spirogyra is a genus of filamentous green algae of the order Zygnematales and there are more than 400 species of Spirogyra in the world. Spirogyra measures approximately 10 to 100μm in width and may stretch centimeters long. Spirogyra is unbranched with cylindrical cells connected end to end in long green filaments.
Budding	Budding is a form of asexual reproduction. The new organism is naturally genetically identical to the primary one (a clone.) When yeast buds, one cell becomes two cells.
Invertebrate	An Invertebrate is an animal without a vertebral column. The group includes 95% of all animal species -- all animals except those in the Chordate subphylum Vertebrata (fish, reptiles, amphibians, birds, and mammals.) Carolus Linnaeus' Systema Naturae divided these animals into only two groups, the Insecta and the now-obsolete vermes (worms.)
Spicules	Spicules are tiny spike-like structures of diverse origin and function found in many organisms, such as the copulatory Spicules of certain nematodes or the grains on the skin of some frogs They provide structural support and deter predators.
Flatworms	The Flatworms, known in scientific literature as Platyhelminthes are a phylum of relatively simple bilaterian, unsegmented, soft-bodied invertebrate animals. Unlike other bilaterians they have no body cavity, and no specialized circulatory and respiratory organs, which restricts them to flattened shapes that allow oxygen and nutrients to pass through their bodies by diffusion. In traditional zoology texts Platyhelminthes are divided into Turbellaria, which are mostly non-parasitic animals such as planarians, and three entirely parasitic groups: Cestoda, Trematoda and Monogenea.

Jellyfish

Jellyfish are free-swimming members of the phylum Cnidaria. They have several different morphologies that represent several different cnidarian classes including the Scyphozoa, Staurozoa, Cubozoa, and Hydrozoa The Jellyfish in these groups are also called, respectively, scyphomedusae, stauromedusae, cubomedusae, and hydromedusae; medusa is another word for Jellyfish.

Medusa

In Greek mythology, Medusa , 'guardian, protectress') was a gorgon, a chthonic female monster; gazing upon her would turn onlookers to stone. She was beheaded by the hero Perseus, who thereafter used her head as a weapon until giving it to the goddess Athena to place on her shield. In classical antiquity and today, the image of the head of Medusa finds expression in the evil-averting device known as the Gorgoneion.

Polyp

A Polyp in zoology is one of two forms found in the phylum Cnidaria, the other being the medusa. Polyp s are approximately cylindrical in shape and elongated at the axis of the body. In solitary Polyp s, the aboral end is attached to the substrate by means of a disc-like holdfast, while in colonies of Polyp s it is connected to other Polyp s, either directly or indirectly.

RNA

Ribonucleic acid (RNA) is a biologically important type of molecule that consists of a long chain of nucleotide units. Each nucleotide consists of a nitrogenous base, a ribose sugar, and a phosphate. RNA is very similar to DNA, but differs in a few important structural details: in the cell, RNA is usually single-stranded, while DNA is usually double-stranded; RNA nucleotides contain ribose while DNA contains deoxyribose (a type of ribose that lacks one oxygen atom); and RNA has the base uracil rather than thymine that is present in DNA.

RNA is transcribed from DNA by enzymes called RNA polymerases and is generally further processed by other enzymes.

Ascaris

Ascaris is a genus of parasitic nematode worms known as the giant intestinal roundworms. One species, A. suum, typically infects pigs, while another, A. lumbricoides, affects human populations, typically in sub-tropical and tropical areas with poor sanitation. A. lumbricoides is the largest intestinal roundworm and is the most common helminth infection of humans worldwide, an infection known as ascariasis.

Ascaris lumbricoides

Ascaris lumbricoides is the member of the Ascaris family responsible for the disease ascariasis.

It can reach a length of up to 35 cm.

Ascaris lumbricoides, or 'roundworm', infections in humans occur when an ingested infective egg releases a larval worm that penetrates the wall of the duodenum and enters the bloodstream.

Hookworm

The Hookworm is a parasitic nematode worm that lives in the small intestine of its host, which may be a mammal such as a dog, cat, or human. Two species of Hookworm s commonly infect humans, Ancylostoma duodenale and Necator americanus. Hookworm s are also bilateral, meaning that if cut in half, the worm would be the exact same on each side.

Cephalopods

The Cephalopods are the mollusc class Cephalopoda characterized by bilateral body symmetry, a prominent head, and a modification of the mollusk foot, a muscular hydrostat, into the form of arms or tentacles. Teuthology, a branch of malacology, is the study of Cephalopods.

The class contains two extant subclasses.

Nereis	Nereis is a genus of polychaete worms in the family Nereidae. It comprises many species, most of which are marine, including the sandworm (Nereis virens) and the common clam worm (Nereis succinea.) Nereis possess setae and parapodia for locomotion.
Annelids	The Annelids, collectively called Annelida , are a large phylum of segmented worms, with over 17,000 modern species including ragworms, earthworms and leeches. They are found in marine environments from tidal zones to hydrothermal vents, in freshwater, and in moist terrestrial environments. Although most textbooks still use the traditional division into polychaetes , oligochaetes (which include earthworms) and leech-like species, research since 1997 has radically changed this scheme, viewing leeches as a sub-group of oligochaetes and oligochaetes as a sub-group of polychaetes.
Earthworm	AcanthodrilidaeCriodrilidaeEudrilidaeGlossoscolecidaeLumbricidaeMegascolecidae Earthworm is the common name for the largest members of Oligochaeta (which is either a class or subclass depending on the author) in the phylum Annelida. In classical systems they were placed in the order Opisthopora, on the basis of the male pores opening posterior to the female pores, even though the internal male segments are anterior to the female. Theoretical cladistic studies have placed them instead in the suborder Lumbricina of the order Haplotaxida, but this may again soon change.
Leeches	Leeches are annelids comprising the subclass Hirudinea. There are freshwater, terrestrial, and marine Leeches. Like the Oligochaeta, they share the presence of a clitellum.
Nephridium	A meta Nephridium (pl. metanephridia) is a type of excretory gland or Nephridium found in many types of invertebrates such as annelids, arthropods and molluscs. It typically consists of a ciliated funnel opening into the body cavity or coelom connected to a duct which may be variously glandularized, folded or expanded (vesiculate) and which typically opens to the organism's exterior.
Pearl	A Pearl is a hard, general spherical object produced within the soft tissue (specifically the mantle) of a living shelled mollusk. Just like the shell of a mollusk, a Pearl is made up of calcium carbonate in minute crystalline form, which has been deposited in concentric layers. The ideal Pearl is perfectly round and smooth, but many other shapes of Pearl s (baroque Pearl s) occur.
Photosynthesis	Photosynthesis is a process that converts carbon dioxide into organic compounds, especially sugars, using the energy from sunlight. Photosynthesis occurs in plants, algae, and many species of Bacteria, but not in Archaea. Photosynthetic organisms are called photoautotrophs, since it allows them to create their own food.
Polychaetes	The Polychaeta or Polychaetes are a class of annelid worms, generally marine. Each body segment has a pair of fleshy protrusions called parapodia that bear many bristles, called chaetae, which are made of chitin. Indeed, Polychaetes are sometimes referred to as bristle worms.
Butterfly	A Butterfly is an insect of the order Lepidoptera. Like all Lepidoptera, butterflies are notable for their unusual life cycle with a larval caterpillar stage, an inactive pupal stage, and a spectacular metamorphosis into a familiar and colourful winged adult form. Most species are day-flying so they regularly attract attention.

Chapter 19. Both Water and Land: Animals

Chitin	$Chitin_n$ is a long-chain polymer of a N-acetylglucosamine, a derivative of glucose, and is found in many places throughout the natural world. It is the main component of the cell walls of fungi, the exoskeletons of arthropods, such as crustaceans and insects, including ants, beetles and butterflies, the radula of mollusks and the beaks of cephalopods, including squid and octopuses. Chitin has also proven useful for several medical and industrial purposes.
Crustaceans	Crustaceans are a very large group of arthropods, comprising almost 52,000 described species , and are usually treated as a subphylum . They include various familiar animals, such as crabs, lobsters, crayfish, shrimp, krill and barnacles. The majority of them are aquatic, living in either marine or fresh water environments, but a few groups have adapted to life on land, such as terrestrial crabs, terrestrial hermit crabs and woodlice.
Exoskeleton	An Exoskeleton is an external skeleton that supports and protects an animal's body, in contrast to the internal endoskeleton of, for example, a human. Some animals, such as the tortoise, have both an endoskeleton and an Exoskeleton In popular usage, many of the larger kinds of Exoskeleton s are known as 'shells'.
Monarch	The Monarch is a milkweed butterfly (subfamily Danainae), in the family Nymphalidae. It is perhaps the best known of all North American butterflies. Since the 19th century, it has been found in New Zealand, and in Australia since 1871 where it is called the Wanderer.
APC	APC (adenomatosis polyposis coli) is a human gene that is classified as a tumor suppressor gene. Tumor suppressor genes prevent the uncontrolled growth of cells that may result in cancerous tumors. The protein made by the APC gene plays a critical role in several cellular processes that determine whether a cell may develop into a tumor.
Arachnid	Arachnid s are a class (Arachnid a) of joint-legged invertebrate animals in the subphylum Chelicerata. All Arachnid s have eight legs, although in some species the front pair may convert to a sensory function. The term Arachnid is from the Greek word Î¬ρÎ¬χνη or arachne, meaning spider, and also referring to the mythological figure Arachne.
Book lung	A Book lung is a type of respiration organ used for atmospheric gas exchange and is found in arachnids, such as scorpions and spiders. Each of these organs is found inside a ventral abdominal cavity and connects with the surroundings through a small opening. Book lung s are not related to the lungs of modern land-dwelling vertebrates.
Centipedes	Centipedes are arthropods belonging to the class Chilopoda and the Subphylum Myriapoda. They are elongated metameric animals with one pair of legs per body segment. A key trait uniting this group is a pair of venom claws or forcipules formed from a modified first appendage.
Horseshoe crab	The Horseshoe crab or Atlantic Horseshoe crab (Limulus polyphemus) is a marine chelicerate arthropod. Despite its name, it is more closely related to spiders, ticks, and scorpions than to crabs. Horseshoe crab s are most commonly found in the Gulf of Mexico and along the northern Atlantic coast of North America.
Krill	Krill are a type of shrimp-like marine invertebrate animal. These small crustaceans are important organisms of the zooplankton, particularly as food for baleen whales, manta rays, whale sharks, crabeater seals, and other seals, and a few seabird species that feed almost exclusively on them. Another name is euphausiids, after their taxonomic order Euphausiacea.

Scorpions

Scorpions are predatory arthropod animals of the order Scorpiones within the class Arachnida. There are about 2,000 species of Scorpions, found widely distributed south of about 49° N, except New Zealand and Antarctica. The northernmost part of the world where Scorpions live in the wild is Sheerness on the Isle of Sheppey in the UK, where a small colony of Euscorpius flavicaudis has been resident since the 1860s.

Spider

Spider s (order Araneae) are air-breathing chelicerate arthropods that have eight legs, and chelicerae modified into fangs that inject venom. Spider s are found world-wide on every continent except for Antarctica, and have become established in nearly every ecological niche with the exception of air and sea colonization. As of 2008, approximately 40,000 Spider species, and 109 families have been recorded by taxonomists.

Insects

Insects are arthropods, having a hard exoskeleton, a three-part body (head, thorax, and abdomen), three pairs of jointed legs, compound eyes, and two antennae. They are the most diverse group of animals on the planet and include approximately 2,200 species of praying mantis, 5,000 dragonfly, 20,000 grasshopper, 82,000 true bug, 120,000 fly, 110,000 bee, wasp, ant and sawfly, 170,000 butterfly and moth, and 360,000 beetle species described to date. The number of extant species is estimated at between six and ten million, with over a million species already described.

Sea star

Sea star s are echinoderms belonging to the class Asteroidea. The names Sea star and 'starfish' are sometimes differentiated, with 'starfish' used in a broader sense to include the closely related brittle stars, which make up the class Ophiuroidea, as well as excluding Sea star s which do not have five arms (have Many arms), such as the sun stars and cushion stars.

Sea star s exhibit a superficially radial symmetry.

Chordates

Chordates are a group of animals that includes the vertebrates, together with several closely related invertebrates. They are united by having, at some time in their life cycle, a notochord, a hollow dorsal nerve cord, pharyngeal slits, an endostyle, and a post-anal tail. The phylum Chordata consists of three subphyla: Urochordata, represented by tunicates; Cephalochordata, represented by lancelets;and Craniata, which includes Vertebrata.

Lancelets

The Lancelets are a group of primitive chordates. They are usually found buried in sand in shallow parts of temperate or tropical seas. In Asia, they are harvested commercially for food for humans and domesticated animals.

No-till farming

No-till farming is a way of growing crops from year to year without disturbing the soil through tillage. No-till increases the amount of water in the soil, decreases erosion, increases the amount and variety of life in and on the soil and it increases herbicide usage.

Producing crops usually involves regular tilling that agitates the soil in various ways, usually with tractor-drawn implements.

Tunicate

Tunicate tunicata is a subphylum of a group of underwater saclike filter feeders with incurrent and excurrent siphons, that are members of the phylum Chordata. Most Tunicate s feed by filtering sea water through pharyngeal slits, but some are sub-marine predators such as the Megalodicopia hians. Like other chordates, Tunicate s have a notochord during their early development, but lack myomeric segmentation throughout the body and tail as adults.

Vertebrates	Vertebrates are members of the subphylum Vertebrata, chordates with backbones or spinal columns. The grouping sometimes includes the hagfish, which have no vertebrae, but are genetically quite closely related to lampreys, which do have vertebrae. For this reason, the sub-phylum is sometimes referred to as 'Craniata', as all members do possess a cranium.
Cell	The Cell is the structural and functional unit of all known living organisms. It is the smallest unit of an organism that is classified as living, and is often called the building block of life. Some organisms, such as most bacteria, are unicellular (consist of a single Cell.)
Horse murders	The Horse murders scandal was a form of insurance fraud in the United States in which expensive horses, many of them show jumpers, were insured against death, accident and then killed to collect the insurance money. It is not known how many horses were murdered between the mid 1970s and the mid-1990s, when a Federal Bureau of Investigation (FBI) investigation brought the horse killings to light, but the number is thought to be well over 50, and may have been as high as 100. In addition, in 1977, the heiress Helen Brach disappeared and was presumed by law enforcement agents to have been murdered by the perpetrators of these crimes, because she threatened to report their criminal activity to authorities; continuing investigations into Brach's death began to uncover the insurance fraud in the 1990s.
Hagfish	Hagfish are marine craniates of the class Myxini, also known as Hyperotreti. Myxini is the only class in the clade Craniata that does not also belong to the subphylum Vertebrata. That is, they are the only animals which have a skull but not a vertebral column.
Terrestrial animals	Terrestrial animals are animals that live predominantly or entirely on land, as compared with aquatic animals, which live predominantly or entirely in the water (e.g., fish, lobsters, octopuses) which rely on a combination of aquatic and terrestrial habitats (e.g., frogs.) Terrestrial animals evolved from marine animals (aquatic animals living in the ocean.) The term terrestrial is also frequently used for species that live primarily on the ground, in contrast to arboreal species, which live primarily in trees.
Ectothermic	Ectothermic refers to organisms that control body temperature through external means. As a result, organisms are dependent on environmental heat sources and have relatively low metabolic rates. For example, many reptiles regulate their body temperature by basking in the sun.
Feathers	Feathers are one of the epidermal growths that form the distinctive outer covering on birds. They are considered the most complex integumentary structures found in vertebrates. They are among the outstanding characteristics that distinguish the extant Aves from other living groups.
Reptile	Reptile s are air-breathing, generally 'cold-blooded' (poikilothermic) amniotes that generally have skin covered in scales or scutes. They are tetrapods and lay amniote eggs, whose embryos are surrounded by the amnion membrane. Modern Reptile s inhabit every continent with the exception of Antarctica, and four living orders are currently recognized: · Crocodilia (crocodiles, gavials, caimans, and alligators): 23 species · Sphenodontia : 2 species · Squamata (lizards, snakes, and amphisbaenids ['worm-lizards']): approximately 7,900 species · Testudines (turtles and tortoises): approximately 300 species

The majority of Reptile species are oviparous (egg-laying) although certain species of squamates are capable of giving live birth. This is achieved, either through ovoviviparity (egg retention), or viviparity

Snake

Snakes, like other reptiles, have a skin covered in scales. Snakes are entirely covered with scales or scutes of various shapes and sizes. Scales protect the body of the Snake, aid it in locomotion, allow moisture to be retained within, alter the surface characteristics such as roughness to aid in camouflage, and in some cases even aid in prey capture (such as Acrochordus.)

Beak

The Beak bill or rostrum is an external anatomical structure of birds which is used for eating and for grooming, manipulating objects, killing prey, probing for food, courtship and feeding young. The term also refers to a similar mouthpart in some monotremes, cephalopods, cetaceans, pufferfishes, turtles, Anuran tadpoles and sirens. Comparison of bird Beak s, displaying different shapes adapted to different feeding methods.

Mammal

Mammal s (formally mammal ia) are a class of vertebrate animals whose females are characterized by the possession of mammary glands while both males and females are characterized by sweat glands, hair, three middle ear bones used in hearing, and a neocortex region in the brain.

mammal s are divided into three main categories depending how they are born. These categories are, monotremes, marsupials and placentals.

Platypus

The Platypus is a semi-aquatic mammal endemic to eastern Australia, including Tasmania. Together with the four species of echidna, it is one of the five extant species of monotremes, the only mammals that lay eggs instead of giving birth to live young. It is the sole living representative of its family (Ornithorhynchidae) and genus (Ornithorhynchus), though a number of related species have been found in the fossil record.

Virginia opossum

The Virginia Opossum, commonly known as the North American Opossum, is the only marsupial found in North America north of the Rio Grande River. A solitary and nocturnal animal about the size of a domestic cat, it is a successful opportunist and is found throughout Central America and North America east of the Rockies from Costa Rica to southern Ontario (it was also introduced to California in 1910, and now occupies much of the Pacific coast); it seems to be still expanding its range northward. Its ancestors evolved in South America, but were enabled to invade North America in the Great American Interchange by the formation of the Isthmus of Panama about 3 million years ago.

Human

A Human is a member of a species of bipedal primates in the family Hominidae . DNA and fossil evidence indicates that modern Human s originated in east Africa about 200,000 years ago. When compared to other animals and primates, Human s have a highly developed brain, capable of abstract reasoning, language, introspection and problem solving.

Biotin

Biotin, also known as vitamin H or B_7, is a water-soluble B-complex vitamin which is composed of an ureido (tetrahydroimidizalone) ring fused with a tetrahydrothiophene ring. A valeric acid substituent is attached to one of the carbon atoms of the tetrahydrothiophene ring. Biotin is a cofactor in the metabolism of fatty acids and leucine, and it plays a role in gluconeogenesis.

Anatomy

The anatomy of spiders is in some aspects similar, but also different from that of other arthropods. The following characteristics are common to all spiders: A body with two segments, eight legs, spinnerets, no chewing parts, no wings, and the presence of chelicerae, which spiders use to hold prey, and in most cases, inject venom. Spiders have non-compound eyes, with most species having eight; the spiders known as Haplogynae may have six or fewer, and certain cave-dwelling spiders may have none at all.

Pitcher plant

Pitcher plant s are carnivorous plants whose prey-trapping mechanism features a deep cavity filled with liquid known as a pitfall trap. It has been widely assumed that the various sorts of pitfall trap evolved from rolled leaves, with selection pressure favouring more deeply cupped leaves over evolutionary time. However, some Pitcher plant genera (such as Nepenthes) are placed within clades consisting mostly of flypaper traps: this indicates that this view may be too simplistic, and some pitchers may have evolved from flypaper traps by loss of mucilage.

Escherichia

Escherichia is a genus of Gram-negative, non-spore forming, facultatively anaerobic, rod-shaped bacteria from the family Enterobacteriaceae. Inhabitants of the gastrointestinal tracts of warm-blooded animals, Escherichia species provide a portion of the microbially-derived vitamin K for their host.

While many Escherichia are harmless commensals, particular strains of some species are human pathogens, and are known as the most common cause of urinary tract infections, significant sources of gastrointestinal disease, ranging from simple diarrhea to dysentery-like conditions, as well as a wide-range of other pathogenic states.

Escherichia coli

Escherichia coli , is a Gram negative bacterium that is commonly found in the lower intestine of warm-blooded organisms . Most E. coli strains are harmless, but some, such as serotype O157:H7, can cause serious food poisoning in humans, and are occasionally responsible for costly product recalls. The harmless strains are part of the normal flora of the gut, and can benefit their hosts by producing vitamin K_2, or by preventing the establishment of pathogenic bacteria within the intestine.

RNA

Ribonucleic acid (RNA) is a biologically important type of molecule that consists of a long chain of nucleotide units. Each nucleotide consists of a nitrogenous base, a ribose sugar, and a phosphate. RNA is very similar to DNA, but differs in a few important structural details: in the cell, RNA is usually single-stranded, while DNA is usually double-stranded; RNA nucleotides contain ribose while DNA contains deoxyribose (a type of ribose that lacks one oxygen atom); and RNA has the base uracil rather than thymine that is present in DNA.

RNA is transcribed from DNA by enzymes called RNA polymerases and is generally further processed by other enzymes.

Shigella

Shigella is a genus of Gram-negative, non-spore forming rod-shaped bacteria closely related to Escherichia coli and Salmonella. The causative agent of human shigellosis, Shigella cause disease in primates, but not in other mammals. It is only naturally found in humans and apes.

Shigella dysenteriae

Shigella dysenteriae is a species of the rod-shaped bacterial genus Shigella. Shigella can cause shigellosis (bacillary dysentery.) Shigellae are Gram-negative, non-spore-forming, facultatively anaerobic, non-motile bacteria.

Bud

In botany, a Bud is an undeveloped or embryonic shoot and normally occurs in the axil of a leaf or at the tip of the stem. Once formed, a Bud may remain for some time in a dormant condition, or it may form a shoot immediately.

The Bud s of many woody plants, especially in temperate or cold climates, are protected by a covering of modified leaves called scales which tightly enclose the more delicate parts of the Bud

Deciduous	Deciduous means falling off at maturity or tending to fall off and is typically used in reference to trees or shrubs that lose their leaves seasonally and to the shedding of other plant structures such as petals after flowering or fruit when ripe. In a more specific sense Deciduous means the dropping of a part that is no longer needed, or falling away after its purpose is finished. In plants it is the result of natural processes.
Evergreen	In botany, an Evergreen plant is a plant having leaves all year round. This contrasts with deciduous plants, which completely lose their foliage for part of the year. Leaf persistence in Evergreen plants may vary from a few months (with new leaves constantly being grown as old ones shed), to several decades (over thirty years in Great Basin Bristlecone Pine Pinus longaeva .)
Root	In vascular plants, the Root is the organ of a plant that typically lies below the surface of the soil. This is not always the case, however, since a Root can also be aerial (growing above the ground) or aerating (growing up above the ground or especially above water.) Furthermore, a stem normally occurring below ground is not exceptional either
Root system	In mathematics, a root system is a configuration of vectors in a Euclidean space satisfying certain geometrical properties. The concept is fundamental in the theory of Lie groups and Lie algebras. Since Lie groups (and some analogues such as algebraic groups) and Lie algebras have become important in many parts of mathematics during the twentieth century, the apparently special nature of root system s belies the number of areas in which they are applied.
Shoot	Shoot s are new plant growth, they can include stems, flowering stems with flower buds, leaves. The new growth from seed germination that grows upward is a Shoot where leaves will develop. In the spring, perennial plant Shoot s are the new growth that grows from the ground in herbaceous plants or the new stem and/or flower growth that grows on woody plants.
Perennial plant	A Perennial plant or perennial is a plant that lives for more than two years. When used by gardeners or horticulturalists, this term applies specifically to perennial herbaceous plants. Scientifically, woody plants like shrubs and trees are also perennial in their habit.
Plants	Plants are living organisms belonging to the kingdom Plantae. They include familiar organisms such as trees, herbs, bushes, grasses, vines, ferns, mosses, and green algae. About 350,000 species of Plants, defined as seed Plants, bryophytes, ferns and fern allies, are estimated to exist currently.
Root hair cell	Root hair cells, the rhizoids of many vascular plants, are tubular outgrowths of trichoblasts, the hair-forming cells on the epidermis of a plant root. That is, root hair cells are lateral extensions of a single cell and only rarely branched, thus invisible to the naked eye. Just prior to the Root hair cell development, there is a point of elevated phosphorylase activity.
Chromosome	A Chromosome is an organized structure of DNA and protein that is found in cells. It is a single piece of coiled DNA containing many genes, regulatory elements and other nucleotide sequences. Chromosome s also contain DNA-bound proteins, which serve to package the DNA and control its functions.

Cotyledon

A Cotyledon is a significant part of the embryo within the seed of a plant. Upon germination, the Cotyledon may become the embryonic first leaves of a seedling. The number of Cotyledon s present is one characteristic used by botanists to classify the flowering plants .

Eudicots

Eudicots and Eudicotyledons are terms introduced by Doyle ' Hotton (1991) to refer to a group of flowering plants that had been called 'tricolpates' or 'non-Magnoliid dicots' by previous authors. The term means, literally, 'true dicotyledons' as it contains the majority of plants that have been considered dicotyledons and have typical dicotyledonous characters. The term 'Eudicots' has been widely adopted to refer to one of the two largest clades of angiosperms, monocots being the other.

Phloem

In vascular plants, Phloem is the living tissue that carries organic nutrients (known as photosynthate), particularly sucrose, a sugar, to all parts of the plant where needed. In trees, the Phloem is the innermost layer of the bark, hence the name, derived from the Greek word φλÏŒος 'bark'. The Phloem is mainly concerned with the transport of soluble organic material made during photosynthesis.

Vascular bundle

A Vascular bundle is a part of the transport system in vascular plants. The transport itself happens in vascular tissue, which exists in two forms: xylem and phloem. Both these tissues are present in a Vascular bundle, which in addition will include supporting and protective tissues.

Xylem

In vascular plants, xylem is one of the two types of transport tissue, phloem being the other. The word 'xylem' is derived from classical Greek ξυλον , 'wood', and indeed the best known xylem tissue is wood, though it is found throughout the plant. Its basic function is to transport water.

APC

APC (adenomatosis polyposis coli) is a human gene that is classified as a tumor suppressor gene. Tumor suppressor genes prevent the uncontrolled growth of cells that may result in cancerous tumors. The protein made by the APC gene plays a critical role in several cellular processes that determine whether a cell may develop into a tumor.

Cambium

In botany the Cambium is a layer or layers of tissue that are the source of cells for secondary growth. There are two types of Cambium

· Cork Cambium
· Vascular Cambium .

Cell

The Cell is the structural and functional unit of all known living organisms. It is the smallest unit of an organism that is classified as living, and is often called the building block of life. Some organisms, such as most bacteria, are unicellular (consist of a single Cell.)

Cork cambium

Cork cambium is a tissue found in many vascular plants as part of the periderm. The Cork cambium is a lateral meristem and is responsible for secondary growth that replaces the epidermis in roots and stems. It is found in woody and many herbaceous dicots, gymnosperms and some monocots, which usually lack secondary growth.

Cuticle

A Cuticle is any of a variety of tough but flexible, non-mineral outer coverings of an organism that provide protection. Cuticle s are non-homologous, differing in their origin, structure and chemical composition. Eponychium is the anatomical term for the human Cuticle /span>

In human anatomy, Cuticle refers to several structures.

Horse murders

The Horse murders scandal was a form of insurance fraud in the United States in which expensive horses, many of them show jumpers, were insured against death, accident and then killed to collect the insurance money. It is not known how many horses were murdered between the mid 1970s and the mid-1990s, when a Federal Bureau of Investigation (FBI) investigation brought the horse killings to light, but the number is thought to be well over 50, and may have been as high as 100. In addition, in 1977, the heiress Helen Brach disappeared and was presumed by law enforcement agents to have been murdered by the perpetrators of these crimes, because she threatened to report their criminal activity to authorities; continuing investigations into Brach's death began to uncover the insurance fraud in the 1990s.

Epidermis

The Epidermis is a single-layered group of cells that covers plants leaves, flowers, roots and stems. It forms a boundary between the plant and the external world. The Epidermis serves several functions, it protects against water loss, regulates gas exchange, secretes metabolic compounds, and (especially in roots) absorbs water and mineral nutrients.

Ground tissue

The types of Ground tissue found in plants develop from Ground tissue meristem and consists of three simple tissues:

· Parenchyma (cells with thin primary walls that retain their protoplasm)
· Collenchyma (cells with thick primary walls that retain their protoplasm)
· Sclerenchyma (cells with lignified secondary walls that have lost their protoplasm at maturity, i.e. are 'dead')

Parenchyma is the most common and versatile Ground tissue. It forms, for example, the cortex and pith of stems, the cortex of roots, the mesophyll of leaves, the pulp of fruits, and the endosperm of seeds. Parenchyma cells are living cells and may remain meristematic at maturity, meaning that they are capable of cell division. They have thin but flexible cellulose cell walls, and are generally polygonal when close-packed, but approximately spherical when isolated from their neighbours.

Light-dependent reactions

The Light-dependent reactions are the first stage of photosynthesis. In this process light energy is converted into chemical energy, in the form of the energy-carriers ATP and NADPH. In the light-independent reactions, the formed NADPH and ATP drive the reduction of CO_2 to more useful organic compounds, such as glucose.

The Light-dependent reactions take place on the thylakoid membrane inside a chloroplast.

Meristem

A Meristem is the tissue in all plants consisting of undifferentiated cells (Meristem atic cells) and found in zones of the plant where growth can take place.

The term Meristem was first used by Karl Wilhelm von Nägeli (1817-1891) from his book 'Beiträge zur Wissenschaftlichen Botanik' in 1858. It is derived from the Greek word 'merizein', meaning to divide in recognition of its inherent function.

Parenchyma

Parenchyma is a term used to describe a bulk of a substance. It is used in different ways in animals and in plants.

The term is New Latin, from Greek parenkhuma, visceral flesh, from parenkhein, to pour in beside : para-, beside + en-, in + khein, to pour.

Suberin

Suberin is a waxy substance found in higher plants. Suberin is a main constituent of cork, and is named after the Cork Oak, Quercus suber.

	Suberin is highly hydrophobic and its main function is to prevent water from penetrating the tissue.
Vascular cambium	The Vascular cambium is a lateral meristem in the vascular tissue of plants. The Vascular cambium is the source of both the secondary xylem (inwards, towards the pith) and the secondary phloem (outwards), and is located between these tissues in the stem and root. A few leaves even have a Vascular cambium.
Spirogyra	Spirogyra is a genus of filamentous green algae of the order Zygnematales and there are more than 400 species of Spirogyra in the world. Spirogyra measures approximately 10 to 100μm in width and may stretch centimeters long. Spirogyra is unbranched with cylindrical cells connected end to end in long green filaments.
Petiole	In botany, the Petiole is the small stalk attaching the leaf blade to the stem. The Petiole usually has the same internal structure as the stem. Outgrowths appearing on each side of the Petiole are called stipules.
Tracheids	Tracheids are elongated cells in the xylem of vascular plants that serve in the transport of water and mineral salts. Tracheids are one of two types of tracheary elements, vessel elements being the other. All tracheary elements develop a thick lignified cell wall, and at maturity the protoplast has broken down and disappeared.
Vessel element	A Vessel element is one of the cell types found in xylem, the water conducting tissue of plants. Vessel element s are typically found in the angiosperms but absent from most gymnosperms such as the conifers. Vessel element s are the building blocks of vessels, which constitute the major part of the water transporting system in the plants where they occur.
Cortex	In botany, the Cortex is the outer of the stem or root of a plant, bounded on the outside by the epidermis and on the inside by the endodermis. It is composed mostly of undifferentiated cells, usually large thin-walled parenchyma cells of the ground tissue system. The outer cortical cells often acquire irregularly thickened cell walls, and are called collenchyma cells.
Pith	Pith is a substance that is found in vascular plants. It consists of soft, spongy parenchyma cells, and is located in the center of the stem in eudicots (both herbaceous and woody) and in the center of the roots in monocots. It is encircled by a ring of xylem (woody tissue), and outside that, a ring of phloem (bark tissue.)
Treatise on Invertebrate Paleontology	The Treatise on Invertebrate Paleontology published by the Geological Society of America and the University of Kansas Press, is a definitive multi-authored work of some 50 volumes, written by more than 300 paleontologists, and covering every phylum, class, order, family, and genus of fossil and extant (still living) invertebrate animals. The prehistoric invertebrates are described as to their taxonomy, morphology, paleoecology, stratigraphic and paleogeographic range. However, genera with no fossil record whatsoever have just a very brief listing.
Bark	Bark is the outermost layers of stems and roots of woody plants. Plants with Bark include trees, woody vines and shrubs. Bark refers to all the tissues outside of the vascular cambium and is a nontechnical term.

Lenticel

A Lenticel is an airy aggregation of cells within the structural surfaces of the stems, roots, and other parts of vascular plants. It functions as a pore, providing a medium for the direct exchange of gasses between the internal tissues and atmosphere, thereby bypassing the periderm, which would otherwise prevent this exchange of gases. The name Lenticel, pronounced with a soft c, derives from its lenticular shape.

Secondary growth

In many vascular plants, Secondary growth is the result of the activity of the vascular cambium. The latter is a meristem that divides to produce secondary xylem cells on the inside of the meristem (the adaxial side) and secondary phloem cells on the outside (the abaxial side.) This growth increases the girth of the plant root or stem, rather than its length, hence the phrase 'secondary thickening'.

Cap

Adenylate CAP is an actin-binding protein that was originally identified as a binding partner for adenylate cyclase. It binds actin monomers and sequesters them from the polymerization process. The yeast ortholog of CAP is called Srv2.

Endodermis

Literally meaning 'inner skin,' Endodermis is the layer of tissue deep in vascular plants.

It is the inner most layers of the cortex,with no intercellular spaces.There are thickenings of special materials around cell which check diffusion of water from xylem to cortex.

In plants, it is a thin layer of parenchyma found in roots, just outside the vascular cylinder.

Pericycle

The Pericycle is a cylinder of parenchyma cells that lies just inside the endodermis and is the outer most part of the stele of plants.

In dicots, it also has the capacity to produce lateral roots. Branch roots arise from this primary meristem tissue.

Root cap

The Root cap is a section of tissue at the tip of a plant root. Root cap s contain statoliths which are involved in gravity perception in plants. If the cap is carefully removed the root will grow randomly.

Adaptation

In ocular physiology, Adaptation is the ability of the eye to adjust to various levels of darkness and light.

The human eye can function from very dark to very bright levels of light -- its sensing capabilities reach across nine orders of magnitude. This means that the brightest and the darkest light signal that the eye can sense are a factor of roughly one thousand million apart.

Micronutrients

Micronutrients are nutrients needed throughout life in small quantities. They are dietary minerals needed by the human body in very small quantities (generally less than 100micrograms/day) as opposed to macrominerals which are required in larger quantities. The Microminerals or trace elements include at least iron, cobalt, chromium, copper, iodine, manganese, selenium, zinc and molybdenum.

Mineral uptake

In plants, Mineral uptake is the process in which minerals enter the cellular material, typically following the same pathway as water. The most normal entrance portal for Mineral uptake is through plant roots.(Roots, 2005) Some mineral ions diffuse in-between the cells. In contrast to water, some minerals are actively taken up by plant cells.

Root nodules

Root nodules occur on the roots of plants that associate with symbiotic bacteria.

Under nitrogen limiting conditions, plants from the pea family Fabaceae form a symbiotic relationship with a host-specific strain of bacteria known as rhizobia.

	Within legume nodules, nitrogen gas from the atmosphere is converted into ammonia, which is then assimilated into amino acids (the building blocks of proteins), nucleotides (the building blocks of DNA and Root nodules A as well as the important energy molecule ATP), and other cellular constituents such as vitamins, flavones, and hormones.
Nutrient	A Nutrient is a chemical that an organism needs to live and grow or a substance used in an organism's metabolism which must be taken in from its environment. Nutrient s are the substances that enrich the body. They build and repair tissues, give heat and energy, and regulate body processes.
Transpiration	Transpiration is the evaporation of water from the aerial parts of plants, especially leaves but also stems, flowers and roots. Leaf surfaces are dotted with openings called stoma that are bordered by guard cells. Collectively the structures are called stomata.

Plants

Plants are living organisms belonging to the kingdom Plantae. They include familiar organisms such as trees, herbs, bushes, grasses, vines, ferns, mosses, and green algae. About 350,000 species of Plants, defined as seed Plants, bryophytes, ferns and fern allies, are estimated to exist currently.

Auxins

Auxins are a class of plant growth substance and morphogens (often called phytohormone or plant hormone.) Auxins play an essential role in coordination of many growth and behavioral processes in the plant life cycle, they and the behavior they played in plant growth was first revealed by a Dutch scientist named Fritz Went (1903-1990.)

Auxins derive their name from the Greek word αυξανω .

Hormone

Hormone s are chemicals released by cells that affect cells in other parts of the body. Only a small amount of Hormone is required to alter cell metabolism. It is essentially a chemical messenger that transports a signal from one cell to another.

APC

APC (adenomatosis polyposis coli) is a human gene that is classified as a tumor suppressor gene. Tumor suppressor genes prevent the uncontrolled growth of cells that may result in cancerous tumors. The protein made by the APC gene plays a critical role in several cellular processes that determine whether a cell may develop into a tumor.

DNA

Deoxyribonucleic acid (DNA) is a nucleic acid that contains the genetic instructions used in the development and functioning of all known living organisms and some viruses. The main role of DNA molecules is the long-term storage of information. DNA is often compared to a set of blueprints or a recipe, or a code, since it contains the instructions needed to construct other components of cells, such as proteins and RNA molecules.

Horse murders

The Horse murders scandal was a form of insurance fraud in the United States in which expensive horses, many of them show jumpers, were insured against death, accident and then killed to collect the insurance money. It is not known how many horses were murdered between the mid 1970s and the mid-1990s, when a Federal Bureau of Investigation (FBI) investigation brought the horse killings to light, but the number is thought to be well over 50, and may have been as high as 100. In addition, in 1977, the heiress Helen Brach disappeared and was presumed by law enforcement agents to have been murdered by the perpetrators of these crimes, because she threatened to report their criminal activity to authorities; continuing investigations into Brach's death began to uncover the insurance fraud in the 1990s.

Dormancy

Dormancy is a period in an organism's life cycle when growth, development, and (in animals) physical activity is temporarily suspended. This minimizes metabolic activity and therefore helps an organism to conserve energy. Dormancy tends to be closely associated with environmental conditions.

Gibberellin

Gibberellin s (GAs) are plant hormones that regulate growth and influence various developmental processes, including stem elongation, germination, dormancy, flowering, sex expression, enzyme induction and leaf and fruit senescence.

disease in rice. It was first isolated in 1935 by Teijiro Yabuta, from fungal strains (Gibberella fujikuroi) provided by Kurosawa.

Seedling

A Seedling is a young plant sporophyte developing out of a plant embryo from a seed. Seedling development starts with germination of the seed. A typical young Seedling consists of three main parts: the radicle (embryonic root), the hypocotyl (embryonic shoot), and the cotyledons

Technology	Technology is a broad concept that deals with an animal species' ethology or behavior of usage and of knowledge of tools and crafts, and how it affects the animal species' ability to control and adapt to its environment. Technology is a term with origins in the Greek 'technologia', 'τεχνολογΓα' -- 'techne', 'τÎχνη' and 'logia', 'λογΓα' ('saying'.) However, a strict definition is elusive; 'Technology' can refer to material objects of use to humanity, such as machines, hardware or utensils, but can also encompass broader themes, including systems, methods of organization, and techniques.
Abscisic acid	Abscisic acid is a plant hormone. It functions in many plant developmental processes, including bud dormancy. ABA was originally believed to be involved in abscission - this is now known only to be the case in a small number of plants.
Abscission	Abscission is the shedding of a body part. It most commonly refers to the process by which a plant intentionally drops one or more of its parts, such as a leaf, fruit, flower or seed, though the term is also used to describe the shedding of a claw by an animal, and is also the word used to describe the separation of daughter cells at the end of cytokinesis, a process that generally begins immediately following mitotic telophase. A plant will abscise a part either to discard a member that is no longer necessary, such as a leaf during autumn, or a flower following fertilisation, or for the purposes of reproduction.
Blood type	A Blood type is a classification of blood based on the presence or absence of inherited antigenic substances on the surface of red blood cells These antigens may be proteins, carbohydrates, glycoproteins depending on the blood group system, and some of these antigens are also present on the surface of other types of cells of various tissues. Several of these red blood cell surface antigens, that stem from one allele, collectively form a blood group system.
Callus	In biological research and biotechnology, a Callus of cells is a mass of undifferentiated cells. In plant biology, Callus cells are those cells that cover a plant wound. A Callus cell culture is usually sustained on gel media, much in the same manner as bacteria are grown.
Cystic fibrosis	Cystic fibrosis is a genetic disorder known to be an inherited disease of the secretory glands, including the glands that make mucus and sweat. The hallmarks of Cystic fibrosis are salty tasting skin, normal appetite but poor growth and poor weight gain, excess mucus production, and coughing/shortness of breath. Males can be infertile due to the condition congenital bilateral absence of the vas deferens.
Cytokinins	Cytokinins are a class of plant growth substances (plant hormones) that promote cell division. In the UK the term 'hormones' is not acceptable. They are primarily involved in cell growth, differentiation, and other physiological processes.
Senescence	Senescence is a process induced by evolution into an organism's genetic make up so that it may live to its healthiest until its reproductive age and die slowly and gradually thereafter. senescence encompasses all of the biological processes of a living organism's approaching an advanced age (i.e., the combination of processes of deterioration which follow the period of development of an organism.) The word senescence is derived from the Latin word senex, meaning 'old man' or 'old age' or 'advanced in age'.

Escherichia	Escherichia is a genus of Gram-negative, non-spore forming, facultatively anaerobic, rod-shaped bacteria from the family Enterobacteriaceae. Inhabitants of the gastrointestinal tracts of warm-blooded animals, Escherichia species provide a portion of the microbially-derived vitamin K for their host. While many Escherichia are harmless commensals, particular strains of some species are human pathogens, and are known as the most common cause of urinary tract infections, significant sources of gastrointestinal disease, ranging from simple diarrhea to dysentery-like conditions, as well as a wide-range of other pathogenic states.
Escherichia coli	Escherichia coli , is a Gram negative bacterium that is commonly found in the lower intestine of warm-blooded organisms . Most E. coli strains are harmless, but some, such as serotype O157:H7, can cause serious food poisoning in humans, and are occasionally responsible for costly product recalls. The harmless strains are part of the normal flora of the gut, and can benefit their hosts by producing vitamin K_2, or by preventing the establishment of pathogenic bacteria within the intestine.
Ethylene	Ethylene is the chemical compound with the formula C_2H_4. It is the simplest alkene. Because it contains a carbon-carbon double bond, Ethylene is called an unsaturated hydrocarbon or an olefin.
Shigella	Shigella is a genus of Gram-negative, non-spore forming rod-shaped bacteria closely related to Escherichia coli and Salmonella. The causative agent of human shigellosis, Shigella cause disease in primates, but not in other mammals. It is only naturally found in humans and apes.
Shigella dysenteriae	Shigella dysenteriae is a species of the rod-shaped bacterial genus Shigella. Shigella can cause shigellosis (bacillary dysentery.) Shigellae are Gram-negative, non-spore-forming, facultatively anaerobic, non-motile bacteria.
Chromosome	A Chromosome is an organized structure of DNA and protein that is found in cells. It is a single piece of coiled DNA containing many genes, regulatory elements and other nucleotide sequences. Chromosome s also contain DNA-bound proteins, which serve to package the DNA and control its functions.
Gravitropism	Gravitropism is a turning or growth movement by a plant or fungus in response to gravity. Charles Darwin was one of the first Europeans to document that roots show positive Gravitropism and stems show negative Gravitropism. That is, roots grow in the direction of gravitational pull (i.e., downward) and stems grow in the opposite direction (i.e., upwards.)
Photoperiodicity	Photoperiodicity is the physiological reaction of organisms to the length of day or night. It occurs in plants and animals. Many flowering plants use a photoreceptor protein, such as phytochrome or cryptochrome, to sense seasonal changes in night length, or photoperiod, which they take as signals to flower.
Phytochrome	Phytochrome is a photoreceptor, a pigment that plants use to detect light. It is sensitive to light in the red and far-red region of the visible spectrum. Many flowering plants use it to regulate the time of flowering based on the length of day and night (photoperiodism) and to set circadian rhythms.
Alagille syndrome	Alagille syndrome is a genetic disorder that affects the liver, heart, and other systems of the body. Problems associated with the disorder generally become evident in infancy or early childhood. The disorder is inherited in an autosomal dominant pattern, and the estimated prevalence of Alagille syndrome is 1 in every 100,000 live births.

Spirogyra	Spirogyra is a genus of filamentous green algae of the order Zygnematales and there are more than 400 species of Spirogyra in the world. Spirogyra measures approximately 10 to 100µm in width and may stretch centimeters long. Spirogyra is unbranched with cylindrical cells connected end to end in long green filaments.
Alternation of phases	The Alternation of phases describes the life cycle of plants, fungi and protists. A multicellular diploid phase alternates with a multicellular haploid phase. The term can be confusing for people familiar only with the life cycle of a typical animal.
Cell	The Cell is the structural and functional unit of all known living organisms. It is the smallest unit of an organism that is classified as living, and is often called the building block of life. Some organisms, such as most bacteria, are unicellular (consist of a single Cell.)
Flower	A Flower sometimes known as a bloom or blossom, is the reproductive structure found in Flower ing plants The biological function of a Flower is to mediate the union of male sperm with female ovum in order to produce seeds. The process begins with pollination, is followed by fertilization, leading to the formation and dispersal of the seeds.
Flowering	A flower, sometimes known as a bloom or blossom, is the reproductive structure found in flowering plants The biological function of a flower is to mediate the union of male sperm with female ovum in order to produce seeds. The process begins with pollination, is followed by fertilization, leading to the formation and dispersal of the seeds.
Flowering plants	The Flowering plants or angiosperms (Angiospermae or Magnoliophyta) are the most widespread group of land plants. The Flowering plants and the gymnosperms are the only extant groups of seed plants. The Flowering plants are distinguished from other seed plants by a series of apomorphies, or derived characteristics.
Gametophyte	In plants and algae that undergo alternation of generations, a Gametophyte is the multicellular structure, or phase, that is haploid, containing a single set of chromosomes: The Gametophyte produces male or female gametes (or both), by a process of cell division called mitosis. The fusion of male and female gametes produces a diploid zygote, which develops by repeated mitotic cell divisions into a multicellular sporophyte. Because the sporophyte is the product of the fusion of two haploid gametes, its cells are diploid, containing two sets of chromosomes.
Leukemia inhibitory factor	Leukemia inhibitory factor an interleukin 6 class cytokine, is a chemical in cells that affects their growth and development. Leukemia inhibitory factor derives its name from its ability to induce the terminal differentiation of myeloid leukaemic cells. Other properties attributed to the cytokine include: the growth promotion and cell differentiation of different types of target cells, influence on bone metabolism, cachexia, neural development, embryogenesis and inflammation.
Microspore	In biology, a Microspore is a small spore as contrasted to the larger megaspore. This combination is found only in heterosperous organisms. Most plants that reproduce by spore without seed only produce one class of spore.
Receptacle	In botany, the Receptacle is the thickened part of a stem from which the flower organs grow. In some accessory fruits, for example in pomes or strawberries, the Receptacle gives rise to the edible part of the fruit.

	In phycology, Receptacle s are structures at the ends of branches of algae mainly in the brown algae or Heterokontophyta in the Order Fucales.
Sporophyte	All land plants, and some algae, have life cycles in which a haploid gametophyte generation alternates with a diploid Sporophyte, the generation of a plant or alga that has a double set of chromosomes. A multicellular Sporophyte generation or phase is present in the life cycle of all land plants and in some green algae. For common flowering plants (Angiosperms), the Sporophyte generation comprises almost their whole life cycle (i.e. whole green plant, roots etc), except phases of small reproductive structures (pollen and ovule.)
Carpel	Carpel s are the building blocks of the gynoecium. If a gynoecium has a single carpel it is called monocarpous. If a gynoecium has multiple, distinct (free, unfused) carpel s, it is apocarpous.
Dioecious	Dioecious species are whose members can produce only one type of gamete; each individual organism belonging to a Dioecious species is distinctly male or female . The majority of animal species are Dioecious. In plant sexuality, there are also Dioecious species.
Ovary	In the flowering plants, an Ovary is a part of the female reproductive organ of the flower or gynoecium. Specifically, it is the part of the carpel which holds the ovule(s) and is located above or below or at the point of connection with the base of the petals and sepals. In this picture of a zucchini the petals and sepals are above the Ovary and such a flower is said to have an inferior Ovary; also referred to as epigynous.
Ovule	Ovule literally means 'small egg.' In seed plants, the Ovule is the structure that gives rise to and contains the female reproductive cells. It consists of three parts: The integuments forming its outer layer, the nucellus (or megasporangium), and the megaspore-derived female gametophyte (or megagametophyte) in its center. The megagametophyte (also called embryo sac in flowering plants) produces the egg cell for fertilization.
Sepal	A Sepal is a part of the flower of angiosperms . Sepal s in most flowers are green and lie under the more conspicuous petals. As a collective unit the Sepal s form a calyx, whereas the collection of petals is called the corolla.
Stamen	The Stamen '>warp') is the male organ of a flower. Each Stamen generally has a stalk called the filament , and, on top of the filament, an anther , and pollen sacs, called microsporangia. The development of the microsporangia and the contained haploid gametophytes, (called pollen-grains) is closely comparable with that of the microsporangia in gymnosperms or heterosporous ferns.
Embryo	An Embryo is a multicellular diploid eukaryote in its earliest stage of development, from the time of first cell division until birth, hatching, or germination. In humans, it is called an Embryo until about eight weeks after fertilization (i.e. ten weeks LMP), and from then it is instead called a fetus. 6 week old excised human Embryo /span> The development of the Embryo is called Embryo genesis.
Pollen	Pollen is a fine to coarse powder consisting of microgametophytes (Pollen grains), which produce the male gametes (sperm cells) of seed plants. A hard coat covering the Pollen grain protects the sperm cells during the process of their movement between the stamens of the flower to the pistil of the next flower. Individual Pollen grains are small enough to require magnification to see detail.

Double fertilization

The Parts of a Flower. double fertilization.

double fertilization is a complex fertilization mechanism that has evolved in flowering plants, known as angiosperms. This process involves the joining of a female gametophyte (embryo sac) with two male gametes (sperm.) It begins when a pollen grain adheres to the stigma of the carpel, the female reproductive structure of a flower.

Pollination

Pollination is the process by which pollen is transferred in plants, thereby enabling fertilisation and sexual reproduction. Pollen grains, which contain the male gametes (sperm) to where the female gamete(s) are contained within the carpel; in gymnosperms the pollen is directly applied to the ovule itself. The receptive part of the carpel is called a stigma in the flowers of angiosperms.

Drosophila

Drosophila has long been a favorite model system for geneticists and developmental biologists studying embryogenesis. The small size, short generation time, and large brood size makes it ideal for genetic studies. Transparent embryos facilitate developmental studies.

Cotyledon

A Cotyledon is a significant part of the embryo within the seed of a plant. Upon germination, the Cotyledon may become the embryonic first leaves of a seedling. The number of Cotyledon s present is one characteristic used by botanists to classify the flowering plants .

Fruit

The term fruit has different meanings dependent on context, and the term is not synonymous in food preparation and biology. fruit s are the means by which flowering plants disseminate seeds, and the presence of seeds indicates that a structure is most likely a fruit though not all seeds come from fruit s.

No single terminology really fits the enormous variety that is found among plant fruit s.

Legume

A Legume is a plant in the family Fabaceae (or Leguminosae), or a fruit of these specific plants. A 'Legume' fruit is a simple dry fruit that develops from a simple carpel and usually dehisces (opens along a seam) on two sides. A common name for this type of fruit is a 'pod', although pod is also applied to a few other fruit types, such as vanilla.

Root

In vascular plants, the Root is the organ of a plant that typically lies below the surface of the soil. This is not always the case, however, since a Root can also be aerial (growing above the ground) or aerating (growing up above the ground or especially above water.) Furthermore, a stem normally occurring below ground is not exceptional either

Seed

A Seed, referred to as a kernel in some plants, is a small embryonic plant enclosed in a covering called the Seed coat, usually with some stored food. It is the product of the ripened ovule of gymnosperm and angiosperm plants which occurs after fertilization and some growth within the mother plant. The formation of the Seed completes the process of reproduction in Seed plants (started with the development of flowers and pollination), with the embryo developed from the zygote and the Seed coat from the integuments of the ovule.

Pome

In botany, a Pome is a specialty type of fruit produced by flowering plants in the subfamily Maloideae of the family Rosaceae.

A Pome is an accessory fruit composed of one or more carpels surrounded by accessory tissue. The accessory tissue is interpreted by some specialists as an extension of the receptacle and is then referred to as 'fruit cortex', and by others as a fused hypanthium or 'torus'; it is the most edible part of this fruit.

Ascaris	Ascaris is a genus of parasitic nematode worms known as the giant intestinal roundworms. One species, A. suum, typically infects pigs, while another, A. lumbricoides, affects human populations, typically in sub-tropical and tropical areas with poor sanitation. A. lumbricoides is the largest intestinal roundworm and is the most common helminth infection of humans worldwide, an infection known as ascariasis.
Ascaris lumbricoides	Ascaris lumbricoides is the member of the Ascaris family responsible for the disease ascariasis. It can reach a length of up to 35 cm. Ascaris lumbricoides, or 'roundworm', infections in humans occur when an ingested infective egg releases a larval worm that penetrates the wall of the duodenum and enters the bloodstream.
Plant propagation	Plant propagation is the process of artificially or naturally distributing plants. Tropical fruit such as avocado also benefit from special seed treatments (specificly invented for that particular tropical fruit) Seeds and spores can be used for reproduction (through eg sowing.) Seeds are typically produced from sexual reproduction within a species, since because genetic recombination has occurred plants grown from seed may have different characteristics to its parents.
Tissue culture	Tissue culture is the growth of tissues and/or cells separate from the organism. This is typically facilitated via use of a liquid, semi-solid, or solid growth medium, such as broth or agar. tissue culture commonly refers to the culture of animal cells and tissues, while the more specific term plant tissue culture is used for plants.
Genetic	Genetics is the study of how living things receive common traits from previous generations. These traits are described by the Genetic information carried by a molecule called DNA. The instructions for constructing and operating an organism are contained in the organism's DNA. Every living thing on earth has DNA in its cells. Genes are the hereditary components of DNA that occupy spots on chromosomes and determine characteristics in an organism.
Genetic engineering	Genetic engineering, recombinant DNA technology, genetic modification/manipulation (GM) and gene splicing are terms that apply to the direct manipulation of an organism's genes. Genetic engineering is different from traditional breeding, where the organism's genes are manipulated indirectly. Genetic engineering uses the techniques of molecular cloning and transformation to alter the structure and characteristics of genes directly.
Blight	Blight refers to a specific symptom affecting plants in response to infection by a pathogenic organism. It is simply a rapid and complete chlorosis, browning, then death of plant tissues such as leaves, branches, twigs, or floral organs. Accordingly, many diseases that primarily exhibit this symptom are called Blight s.
Genetically modified plants	Genetically modified plants are genetically engineered to contain one or more genes of another species. The aim is to introduce a new trait to the plant species which does not occur naturally in this species, for example resistance to certain pests, diseases or environmental conditions, or the production of a certain nutrient or pharmaceutical agent. Genetically modified plants are often called 'transgenic plants', as they contain one or more trans genes from other organisms, however, this term also includes plants in which the trans gene was integrated by naturally occurring processes.
Potassium	Potassium is a chemical element. It has the symbol K , atomic number 19, and atomic mass 39.0983. Potassium was first isolated from potash.

Transgenic plants	Transgenic plants possess a gene or genes that have been transferred from a different species. Although DNA of another species can be integrated in a plant genome by natural processes, the term 'Transgenic plants' refers to plants created in a laboratory using recombinant DNA technology. The aim is to design plants with specific characteristics by artificial insertion of genes from other species or sometimes entirely different kingdoms.
Chestnut	Chestnut (some species called chinkapin or chinquapin) is a genus of eight or nine species of deciduous trees and shrubs in the beech family Fagaceae, native to temperate regions of the Northern Hemisphere. The name also refers to the edible nuts they produce. The Chestnut belongs to the same Fagaceae family as the Oak and Beech.
Tobacco mosaic virus	Tobacco mosaic virus is an RNA virus that infects plants, especially tobacco and other members of the family Solanaceae. The infection causes characteristic patterns (mottling and discoloration) on the leaves (hence the name.) tobacco mosaic virus was the first virus to be discovered.

Horse murders

The Horse murders scandal was a form of insurance fraud in the United States in which expensive horses, many of them show jumpers, were insured against death, accident and then killed to collect the insurance money. It is not known how many horses were murdered between the mid 1970s and the mid-1990s, when a Federal Bureau of Investigation (FBI) investigation brought the horse killings to light, but the number is thought to be well over 50, and may have been as high as 100. In addition, in 1977, the heiress Helen Brach disappeared and was presumed by law enforcement agents to have been murdered by the perpetrators of these crimes, because she threatened to report their criminal activity to authorities; continuing investigations into Brach's death began to uncover the insurance fraud in the 1990s.

Nereis

Nereis is a genus of polychaete worms in the family Nereidae. It comprises many species, most of which are marine, including the sandworm (Nereis virens) and the common clam worm (Nereis succinea.) Nereis possess setae and parapodia for locomotion.

Spirogyra

Spirogyra is a genus of filamentous green algae of the order Zygnematales and there are more than 400 species of Spirogyra in the world. Spirogyra measures approximately 10 to 100µm in width and may stretch centimeters long.

Spirogyra is unbranched with cylindrical cells connected end to end in long green filaments.

Columnar

In biology, Columnar refers to the shape of epithelial cells that are taller than they are wide. Form follows function in biology, and Columnar morphorphology hints at the functions of the cell. Columnar cells are important in absorption and movement of mucus.

Squamous epithelium

In anatomy, Squamous epithelium is an epithelium characterised by its most superficial layer consisting of flat, scale-like cells called squamous cell. Epithelium may possess only one layer of these cells, in which case it is referred to as simple Squamous epithelium; or it may possess multiple layers, referred to then as stratified Squamous epithelium. Both types perform differing functions, ranging from nutrient exchange to protection.

Decomposers

Decomposers are organisms that consume dead or decaying organisms, and, in doing so, carry out the natural process of decomposition. Like herbivores and predators, Decomposers are heterotrophic, meaning that they use organic substrates to get their energy, carbon and nutrients for growth and development. Decomposers use deceased organisms and non-living organic compounds as their food source.

Fibroblast

A Fibroblast is a type of cell that synthesizes the extracellular matrix and collagen, the structural framework (stroma) for animal tissues, and play a critical role in wound healing. They are the most common cells of connective tissue in animals.

Fibroblast s and fibrocytes are two states of the same cells, the former being the activated state, the latter the less active state, concerned with maintenance.

Pigment

A pigment is the material that changes the color of light it reflects as the result of selective color absorption. This physical process differs from fluorescence, phosphorescence, and other forms of luminescence, in which the material itself emits light.

Many materials selectively absorb certain wavelengths of light.

Blood cell

A Blood cell is any cell of any type normally found in blood. In mammals, these fall into three general categories:

· Red Blood cell s - Erythrocytes

· White Blood cell s- Leukocytes

· Platelets - Thrombocytes red and white human Blood cell s as seen under a microscope using a blue slide stain Together, these three kinds of Blood cell s sum up for a total 45% of blood tissue

.

Operon

An Operon is a functioning unit of key nucleotide sequences of DNA including an operator, a common promoter, and one or more structural genes, which is controlled as a unit to produce messenger RNA (mRNA), in the process of transcription by an RNA polymerase. A typical Operon.

The term "Operon" was first proposed in a short paper in the Proceedings of the French Academy of Science in 1960. From this paper, the so-called general theory of the Operon was developed.

Red blood cells

Red blood cells are the most common type of blood cell and the vertebrate body's principal means of delivering oxygen to the body tissues via the blood. They take up oxygen in the lungs or gills and release it while squeezing through the body's capillaries. The cells are filled with hemoglobin, a biomolecule that can bind to oxygen.

Virus

A virus is a microscopic infectious agent that can reproduce only inside a host cell. virus es infect all types of organisms: from animals and plants, to bacteria and archaea. Since the initial discovery of tobacco mosaic virus by Martinus Beijerinck in 1898, more than 5,000 types of virus have been described in detail, although most types of virus remain undiscovered.

White

White was the first sex-linked mutation ever discovered in Drosophila melanogaster. In 1910 Thomas Hunt Morgan, (or, legend has it, his wife) collected a single male White-eyed mutant from a population of Drosophila melanogaster fruit flies, which usually have bright red eyes. Upon breeding this male with wild-type female flies he found that the offspring did not conform to the expectations of Mendelian inheritance.

White blood cells

White blood cells are cells of the immune system defending the body against both infectious disease and foreign materials. Five different and diverse types of leukocytes exist, but they are all produced and derived from a multipotent cell in the bone marrow known as a hematopoietic stem cell. Leukocytes are found throughout the body, including the blood and lymphatic system.

Shigella

Shigella is a genus of Gram-negative, non-spore forming rod-shaped bacteria closely related to Escherichia coli and Salmonella. The causative agent of human shigellosis, Shigella cause disease in primates, but not in other mammals. It is only naturally found in humans and apes.

Shigella dysenteriae

Shigella dysenteriae is a species of the rod-shaped bacterial genus Shigella. Shigella can cause shigellosis (bacillary dysentery.) Shigellae are Gram-negative, non-spore-forming, facultatively anaerobic, non-motile bacteria.

Axon

An Axon or nerve fiber is a long, slender projection of a nerve cell that conducts electrical impulses away from the neuron's cell body or soma.

An Axon is one of two types of protoplasmic protrusions that extrude from the cell body of a neuron, the other type being dendrites. Axon s are distinguished from dendrites by several features, including shape (dendrites often taper while Axon s usually maintain a constant radius), length (dendrites are restricted to a small region around the cell body while Axon s can be much longer), and function (dendrites usually receive signals while Axon s usually transmit them.)

Dendrites

Dendrites are the branched projections of a neuron that act to conduct the electrochemical stimulation received from other neural cells to the cell body of the neuron from which the Dendrites project. Electrical stimulation is transmitted onto Dendrites by upstream neurons via synapses which are located at various points throughout the dendritic arbor. Dendrites play a critical role in integrating these synaptic inputs and in determining the extent to which action potentials are produced by the neuron.

Immune system

An immune system is a collection of biological processes within an organism that protects against disease by identifying and killing pathogens and tumour cells. It detects a wide variety of agents, from viruses to parasitic worms, and needs to distinguish them from the organism's own healthy cells and tissues in order to function properly. Detection is complicated as pathogens can evolve rapidly, producing adaptations that avoid the immune system and allow the pathogens to successfully infect their hosts.

Interstitial fluid

Interstitial fluid is a solution which bathes and surrounds the cells of multicellular animals. It is the main component of the extracellular fluid, which also includes plasma and transcellular fluid. The Interstitial fluid is found in the interstitial spaces, also known as the tissue spaces.

Endocrine system

The Endocrine system is a system of glands that involve the release of extracellular signaling molecules known as hormones. The Endocrine system is instrumental in regulating metabolism, growth, development and puberty, and tissue function and also plays a part in determining mood. The field of study that deals with disorders of endocrine glands is endocrinology, a branch of the wider field of internal medicine.

RNA

Ribonucleic acid (RNA) is a biologically important type of molecule that consists of a long chain of nucleotide units. Each nucleotide consists of a nitrogenous base, a ribose sugar, and a phosphate. RNA is very similar to DNA, but differs in a few important structural details: in the cell, RNA is usually single-stranded, while DNA is usually double-stranded; RNA nucleotides contain ribose while DNA contains deoxyribose (a type of ribose that lacks one oxygen atom); and RNA has the base uracil rather than thymine that is present in DNA.

RNA is transcribed from DNA by enzymes called RNA polymerases and is generally further processed by other enzymes.

Helicobacter	Helicobacter is a genus of Gram-negative bacteria possessing a characteristic helix shape. They were initially considered to be members of the Campylobacter genus, but since 1989 they have been grouped in their own genus. Some species have been found living in the lining of the upper gastrointestinal tract, as well as the liver of mammals and some birds..
Helicobacter pylori	Helicobacter pylori is a Gram-negative, microaerophilic bacterium that inhabits various areas of the stomach and duodenum. It causes a chronic low-level inflammation of the stomach lining and is strongly linked to the development of duodenal and gastric ulcers and stomach cancer. Over 80% of individuals infected with the bacterium are asymptomatic.
Shigella	Shigella is a genus of Gram-negative, non-spore forming rod-shaped bacteria closely related to Escherichia coli and Salmonella. The causative agent of human shigellosis, Shigella cause disease in primates, but not in other mammals. It is only naturally found in humans and apes.
Shigella dysenteriae	Shigella dysenteriae is a species of the rod-shaped bacterial genus Shigella. Shigella can cause shigellosis (bacillary dysentery.) Shigellae are Gram-negative, non-spore-forming, facultatively anaerobic, non-motile bacteria.
Hemocoel	A Hemocoel is a series of spaces between the organs of organisms with open circulatory systems, like most arthropods and mollusks. A combination of blood, lymph, and interstitial fluid called hemolymph circulates through the Hemocoel. The term is also used to describe the space in the connective tissue compartment derived from the embryonic blastocoel with contributions from mesoderm, filled with blood.
Sinus	In botany, a Sinus is a space or indentation, usually on a leaf, between two lobes or teeth that does not break the continuity of the structure.
Protein	Protein s are organic compounds made of amino acids arranged in a linear chain. The amino acids in a polymer chain are joined together by the peptide bonds between the carboxyl and amino groups of adjacent amino acid residues. The sequence of amino acids in a protein is defined by the sequence of a gene, which is encoded in the genetic code.
Vertebrates	Vertebrates are members of the subphylum Vertebrata, chordates with backbones or spinal columns. The grouping sometimes includes the hagfish, which have no vertebrae, but are genetically quite closely related to lampreys, which do have vertebrae. For this reason, the sub-phylum is sometimes referred to as 'Craniata', as all members do possess a cranium.
ATP synthase	An ATP synthase is a general term for an enzyme that can synthesize adenosine triphosphate (ATP) from adenosine diphosphate (ADP) and inorganic phosphate by using some form of energy. This energy is often in the form of protons moving down an electrochemical gradient, such as from the lumen into the stroma of chloroplasts or from the inter-membrane space into the matrix in mitochondria. The overall reaction sequence is: $ADP + P_i \rightarrow ATP$ These enzymes are of crucial importance in almost all organisms, because ATP is the common 'energy currency' of cells.

Human

A Human is a member of a species of bipedal primates in the family Hominidae . DNA and fossil evidence indicates that modern Human s originated in east Africa about 200,000 years ago. When compared to other animals and primates, Human s have a highly developed brain, capable of abstract reasoning, language, introspection and problem solving.

Septum

Septa (singular Septum) are thin walls or partitions between the internal chambers (camerae) of the shell of a cephalopod, namely nautiloids or ammonoids.

As the creature grows, its body moves forward in the shell to a new living chamber, secreting septa behind it. This adds new chambers to the shell, which can be clearly seen in cross-sections of the shell of the living nautilus, or in ammonoid and nautiloid fossils.

Valve

A Valve is a device that regulates the flow of a fluid (gases, fluidized solids, slurries closing but are usually discussed as a separate category.

Valve s are also found in the human body.

Edward

The Edward mango is a monoembryonic mango cultivar grown predominantly in Florida. It is considered by many to be among the finest tasting mangoes in the world; however, its poor yields have restrained the Edward from developing into a commercially significant variety.

The Edward was first propagated in the 1920s by Edward Simmonds of the Plant Introduction Garden in Miami, Florida and is believed to be a hybrid cross of Haden and Carabao mango cultivars.

Pulse

In medicine, a person's Pulse is the arterial palpation of a heartbeat. It can be palpated in any place that allows for an artery to be compressed against a bone, such as at the neck (carotid artery), at the wrist (radial artery), behind the knee (popliteal artery), on the inside of the elbow (brachial artery), and near the ankle joint (posterior tibial artery.) The Pulse rate can also be measured by measuring the heart beats directly (the apical Pulse)

Armadillo

Armadillo s are small placental mammals, known for having a leathery armor shell. The Dasypodidae are the only surviving family in the order Cingulata, part of the superorder Xenarthra along with the anteaters and sloths. The word Armadillo is Spanish for 'little armored one'.

Blood cell

A Blood cell is any cell of any type normally found in blood. In mammals, these fall into three general categories:

· Red Blood cell s - Erythrocytes

· White Blood cell s- Leukocytes

· Platelets - Thrombocytes red and white human Blood cell s as seen under a microscope using a blue slide stain

Together, these three kinds of Blood cell s sum up for a total 45% of blood tissue

.

Erythropoietin

Erythropoietin is a glycoprotein hormone that controls erythropoiesis it is produced by the peritubular capillary endothelial cells in the kidney, and is the hormone that regulates red blood cell production.

Hemoglobin	Hemoglobin is the iron-containing oxygen-transport metalloprotein in the red blood cells of vertebrates, and the tissues of some invertebrates. In mammals, the protein makes up about 97% of the red blood cell's dry content, and around 35% of the total content . Hemoglobin transports oxygen from the lungs or gills to the rest of the body where it releases the oxygen for cell use.
Red blood cells	Red blood cells are the most common type of blood cell and the vertebrate body's principal means of delivering oxygen to the body tissues via the blood. They take up oxygen in the lungs or gills and release it while squeezing through the body's capillaries. The cells are filled with hemoglobin, a biomolecule that can bind to oxygen.
Nereis	Nereis is a genus of polychaete worms in the family Nereidae. It comprises many species, most of which are marine, including the sandworm (Nereis virens) and the common clam worm (Nereis succinea.) Nereis possess setae and parapodia for locomotion.
Macrophages	Macrophages are white blood cells within tissues, produced by the division of monocytes. Human Macrophages are about 21 micrometres in diameter. Monocytes and Macrophages are phagocytes, acting in both non-specific defense as well as to help initiate specific defense mechanisms (or adaptive immunity) of vertebrate animals.
Monocyte	Monocyte is a type of white blood cell, part of the human body's immune system. Monocyte s have two main functions in the immune system: (1) replenish resident macrophages and dendritic cells under normal states, and (2) in response to inflammation signals, Monocyte s can move quickly (approx. 8-12 hours) to sites of infection in the tissues and divide/differentiate into macrophages and dendritic cells to elicit an immune response.
Virus	A virus is a microscopic infectious agent that can reproduce only inside a host cell. virus es infect all types of organisms: from animals and plants, to bacteria and archaea. Since the initial discovery of tobacco mosaic virus by Martinus Beijerinck in 1898, more than 5,000 types of virus have been described in detail, although most types of virus remain undiscovered.
White	White was the first sex-linked mutation ever discovered in Drosophila melanogaster. In 1910 Thomas Hunt Morgan, (or, legend has it, his wife) collected a single male White-eyed mutant from a population of Drosophila melanogaster fruit flies, which usually have bright red eyes. Upon breeding this male with wild-type female flies he found that the offspring did not conform to the expectations of Mendelian inheritance.
White blood cells	White blood cells are cells of the immune system defending the body against both infectious disease and foreign materials. Five different and diverse types of leukocytes exist, but they are all produced and derived from a multipotent cell in the bone marrow known as a hematopoietic stem cell. Leukocytes are found throughout the body, including the blood and lymphatic system.
Vitamin E	Vitamin E is the collective name for a set of 8 related α-, β-, γ-, and δ-tocopherols and the corresponding four tocotrienols, which are fat-soluble vitamins with antioxidant properties. Of these, α-tocopherol (also written as alpha-tocopherol) has been most studied as it has the highest bioavailability. It has been claimed that α-tocopherol is the most important lipid-soluble antioxidant, and that it protects cell membranes from oxidation by reacting with lipid radicals produced in the lipid peroxidation chain reaction.

Vitamin K	Vitamin K denotes a group of lipophilic, hydrophobic vitamins that are needed for the posttranslational modification of certain proteins, mostly required for blood coagulation. Chemically they are 2-methyl-1,4-naphthoquinone derivatives. Vitamin K_1 is also known as phylloquinone or phytomenadione .

Magnesium

Magnesium is an essential element in biological systems. Magnesium occurs typically as the Mg^{2+} ion. It is an essential mineral nutrient for life and is present in every cell type in every organism.

Maintenance

Maintenance of an organism is the collection of processes to stay alive, excluding production processes. The Dynamic Energy Budget theory delineates two classes

· Somatic Maintenance. This comprises the turnover of structural mass (mainly proteins), the Maintenance of concentration gradients of metabolites across membranes, activity
· Maturity Maintenance. This comprises the Maintenance of defence systems (such as the immune system), the preparation of the body for reproduction.

The theory assumes that maturity Maintenance costs can be reduced more easily during starvation than somatic Maintenance costs. Under extreme starvation conditions, somatic Maintenance costs are paid from structural mass, which causes shrinking.Some organism manage to switch to the turpor state under starvation conditions, and reduce their Maintenance costs.

NADPH

Nicotinamide adenine dinucleotide phosphate ($NADP^+$, in older notation triphosphopyridine nucleotide, TPN) is used in anabolic reactions, such as lipid and nucleic acid synthesis, which require NADPH as a reducing agent.

NADPH is the reduced form of $NADP^+$, and $NADP^+$ is the oxidized form of NADPH. NADP+ differs from NAD+ by the presence in NADP+ of an additional phosphate group on the 2' position of the ribose ring that carries the adenine moiety. In chloroplasts, NADP is reduced by ferredoxin-NADP+ reductase in the last step of the electron chain of the light reactions of photosynthesis.

Shigella

Shigella is a genus of Gram-negative, non-spore forming rod-shaped bacteria closely related to Escherichia coli and Salmonella. The causative agent of human shigellosis, Shigella cause disease in primates, but not in other mammals. It is only naturally found in humans and apes.

Shigella dysenteriae

Shigella dysenteriae is a species of the rod-shaped bacterial genus Shigella. Shigella can cause shigellosis (bacillary dysentery.) Shigellae are Gram-negative, non-spore-forming, facultatively anaerobic, non-motile bacteria.

Globular protein

Globular protein s comprising 'globe'-like proteins that are more or less soluble in aqueous solutions (where they form colloidal solutions.) This main characteristic helps distinguishing them from fibrous proteins (the other class), which are practically insoluble.

The term globin can refer more specifically to proteins including the globin fold.

Human

A Human is a member of a species of bipedal primates in the family Hominidae . DNA and fossil evidence indicates that modern Human s originated in east Africa about 200,000 years ago. When compared to other animals and primates, Human s have a highly developed brain, capable of abstract reasoning, language, introspection and problem solving.

Photosynthesis

Photosynthesis is a process that converts carbon dioxide into organic compounds, especially sugars, using the energy from sunlight. Photosynthesis occurs in plants, algae, and many species of Bacteria, but not in Archaea. Photosynthetic organisms are called photoautotrophs, since it allows them to create their own food.

Sinus
In botany, a Sinus is a space or indentation, usually on a leaf, between two lobes or teeth that does not break the continuity of the structure.

Alagille syndrome
Alagille syndrome is a genetic disorder that affects the liver, heart, and other systems of the body. Problems associated with the disorder generally become evident in infancy or early childhood. The disorder is inherited in an autosomal dominant pattern, and the estimated prevalence of Alagille syndrome is 1 in every 100,000 live births.

Ascaris
Ascaris is a genus of parasitic nematode worms known as the giant intestinal roundworms. One species, A. suum, typically infects pigs, while another, A. lumbricoides, affects human populations, typically in sub-tropical and tropical areas with poor sanitation. A. lumbricoides is the largest intestinal roundworm and is the most common helminth infection of humans worldwide, an infection known as ascariasis.

Ascaris lumbricoides
Ascaris lumbricoides is the member of the Ascaris family responsible for the disease ascariasis.

It can reach a length of up to 35 cm.

Ascaris lumbricoides, or 'roundworm', infections in humans occur when an ingested infective egg releases a larval worm that penetrates the wall of the duodenum and enters the bloodstream.

Brittle stars
Brittle stars are echinoderms, closely related to sea stars. They crawl across the seafloor using their flexible arms for locomotion. The ophiuroids generally have five long slender, whip-like arms which may reach up to 60 centimeters (2 feet) in length on the largest specimens.

Gill
A Gill is an anatomical structure found in many aquatic organisms. It is a respiration organ whose function is the extraction of oxygen from water and the excretion of carbon dioxide. The microscopic structure of a Gill is such that it presents a very large surface area to the external environment.

Membrane
A Membrane is a layer of material which serves as a selective barrier between two phases and remains impermeable to specific particles, molecules, or substances when exposed to the action of a driving force. Some components are allowed passage by the Membrane into a permeate stream, whereas others are retained by it and accumulate in the retentate stream.

Membrane s can be of various thickness, with homogeneous or heterogeneous structure.

Helicobacter
Helicobacter is a genus of Gram-negative bacteria possessing a characteristic helix shape. They were initially considered to be members of the Campylobacter genus, but since 1989 they have been grouped in their own genus.

Some species have been found living in the lining of the upper gastrointestinal tract, as well as the liver of mammals and some birds..

Helicobacter pylori
Helicobacter pylori is a Gram-negative, microaerophilic bacterium that inhabits various areas of the stomach and duodenum. It causes a chronic low-level inflammation of the stomach lining and is strongly linked to the development of duodenal and gastric ulcers and stomach cancer. Over 80% of individuals infected with the bacterium are asymptomatic.

Heme

A Heme or haem is a prosthetic group that consists of an iron atom contained in the center of a large heterocyclic organic ring called a porphyrin. Not all porphyrins contain iron, but a substantial fraction of porphyrin-containing metalloproteins have Heme as their prosthetic group; these are known as hemoproteins. The histidine bound Heme group of succinate dehydrogenase, an electron carrier in the mitochondrial electron transfer chain.

Ion

An ion is an atom or molecule where the total number of electrons is not equal to the total number of protons, giving it a net positive or negative electrical charge.

Since protons are positively charged and electrons are negatively charged, if there are more electrons than protons, the atom or molecule will be negatively charged. This is called an an ion , from the Greek á¼€vÎ¬ , meaning 'up'.

Cortex

In botany, the Cortex is the outer of the stem or root of a plant, bounded on the outside by the epidermis and on the inside by the endodermis. It is composed mostly of undifferentiated cells, usually large thin-walled parenchyma cells of the ground tissue system. The outer cortical cells often acquire irregularly thickened cell walls, and are called collenchyma cells.

Urine

Urine is a liquid waste product of the body secreted by the kidneys by a process of filtration from blood called urination and excreted through the urethra. Cellular metabolism generates numerous waste compounds, many rich in nitrogen, that require elimination from the bloodstream. This waste is eventually expelled from the body in a process known as micturition, the primary method for excreting water-soluble chemicals from the body.

Capsule

In botany a Capsule is a type of simple, dry fruit produced by many species of flowering plants. A Capsule is a dehiscent structure composed of two or more carpels, that, at maturity, split apart (dehisce) to release the seeds within. In some capsules, the split occurs between carpels, and in others each carpel splits open.In yet others, seeds are released through openings or pores that form in the Capsule.

Secretion

Secretion is the process of elaborating and releasing chemicals from a cell, a secreted chemical substance or amount of substance. In contrast to excretion, the substance may have a certain function, rather than being a waste product.

Secretion in bacterial species means the transport or translocation of effector molecules for example proteins, enzymes or toxins (such as cholera toxin in pathogenic bacteria for example Vibrio cholerae) from across the interior (cytoplasm or cytosol) of a bacterial cell to its exterior.

Acetabularia

Acetabularia is a genus of green algae, specifically of the Polyphysaceae family, Typically found in subtropical waters, Acetabularia is a single-cell organism, but gigantic in size and complex in form, making it an excellent model organism for studying cell biology. In form, the mature Acetabularia resembles the round leaves of a nasturtium, being 0.5 to 10 cm tall and having three anatomical parts: a bottom rhizoid that resembles a set of short roots; a long stalk in the middle; and a top umbrella of branches that may fuse into a cap. The single nucleus of Acetabularia is located in the rhizoid, and allows the cell to regenerate completely if its cap is removed.

Salmonella

Salmonella is a genus of rod-shaped, Gram-negative, non-spore forming, predominantly motile enterobacteria with diameters around 0.7 to 1.5 Âµm, lengths from 2 to 5 Âµm, and flagella which project in all directions (i.e. peritrichous.) They are chemoorganotrophs, obtaining their energy from oxidation and reduction reactions using organic sources and are facultative anaerobes; most species produce hydrogen sulfide, which can readily be detected by growing them on media containing ferrous sulfate, such as TSI. Most isolates exist in two phases; phase I is the motile phase and phase II the non-motile phase. Cultures that are non-motile upon primary culture may be swithched to the motile phase using a Craigie tube.

Acetylcholine

The chemical compound Acetylcholine is a neurotransmitter in both the peripheral nervous system (PNS) and central nervous system (CNS) in many organisms including humans. Acetylcholine is one of many neurotransmitters in the autonomic nervous system (ANS) and the only neurotransmitter used in the motor division of the somatic nervous system. (Sensory neurons use glutamate and various peptides at their synapses.)

Dialysis

In biochemistry, Dialysis is the process of separating molecules in solution by the difference in their rates of diffusion through a semipermeable membrane, such as Dialysis tubing.

Dialysis is a common laboratory technique, and operates on the same principle as medical Dialysis. Typically a solution of several types of molecules is placed into a semipermeable Dialysis bag, such as a cellulose membrane with pores, and the bag is sealed.

Body plan

A Body plan is essentially the blueprint for the way the body of an organism is laid out. An organism's symmetry, its number of body segments and number of limbs are all aspects of its Body plan One of the key issues of developmental biology is the evolution of Body plan s as different as those of a starfish, a fern, or a mammal, from a common biological heritage, and in particular how radical changes in Body plan s have occurred over geological time.

Coelom

The Coelom is a fluid filled cavity formed within the mesoderm. Coelom s developed in triploblasts but were subsequently lost in several lineages. Loss of Coelom is correlated with reduction in body size.

Amylase

An Amylase is an enzyme that breaks starch down into sugar. Amylase is present in human saliva, where it begins the chemical process of digestion. Foods that contain much starch but little sugar, such as rice and potato, taste slightly sweet as they are chewed because Amylase turns some of their starch into sugar in the mouth.

Herbivore

Herbivory is a form of predation in which an organism consumes principally autotrophs such as plants, algae and photosynthesizing bacteria. By that definition, many fungi, some bacteria, many animals, some protists and a small number of parasitic plants can be considered herbivore s. However, herbivory is generally restricted to animals eating plants.

Escherichia

Escherichia is a genus of Gram-negative, non-spore forming, facultatively anaerobic, rod-shaped bacteria from the family Enterobacteriaceae. Inhabitants of the gastrointestinal tracts of warm-blooded animals, Escherichia species provide a portion of the microbially-derived vitamin K for their host.

While many Escherichia are harmless commensals, particular strains of some species are human pathogens, and are known as the most common cause of urinary tract infections, significant sources of gastrointestinal disease, ranging from simple diarrhea to dysentery-like conditions, as well as a wide-range of other pathogenic states.

Escherichia coli

Escherichia coli , is a Gram negative bacterium that is commonly found in the lower intestine of warm-blooded organisms . Most E. coli strains are harmless, but some, such as serotype O157:H7, can cause serious food poisoning in humans, and are occasionally responsible for costly product recalls. The harmless strains are part of the normal flora of the gut, and can benefit their hosts by producing vitamin K_2, or by preventing the establishment of pathogenic bacteria within the intestine.

Horse murders

The Horse murders scandal was a form of insurance fraud in the United States in which expensive horses, many of them show jumpers, were insured against death, accident and then killed to collect the insurance money. It is not known how many horses were murdered between the mid 1970s and the mid-1990s, when a Federal Bureau of Investigation (FBI) investigation brought the horse killings to light, but the number is thought to be well over 50, and may have been as high as 100. In addition, in 1977, the heiress Helen Brach disappeared and was presumed by law enforcement agents to have been murdered by the perpetrators of these crimes, because she threatened to report their criminal activity to authorities; continuing investigations into Brach's death began to uncover the insurance fraud in the 1990s.

Bile

Bile or gall is a bitter yellow, blue and green fluid secreted by hepatocytes from the liver of most vertebrates. In many species, Bile is stored in the gallbladder between meals and upon eating is discharged into the duodenum where the Bile aids the process of digestion of lipids by emulsification.

Bile has various components, some of which are produced by hepatocytes in the liver.

Enzymes

Enzymes are biomolecules that catalyze (i.e., increase the rates of) chemical reactions. Nearly all known Enzymes are proteins. However, certain RNA molecules can be effective biocatalysts too.

Ammagnostidae

Ammagnostidae is a family of trilobites in the superfamily Agnostoidea, which is part of the order Agnostida 2 or 3 segmented trilobites. It has four genera:

Ammagnostus Öpik, 1967

· Ammagnostus psammius Öpik, 1967 (Type)
· Ammagnostus bassus (Öpik, 1967) Guo and Luo, 1996
· Ammagnostus bella Guo and Luo, 1996
· Ammagnostus beltensis (Lochman, 1944) Robinson, 1988
· Ammagnostus cryptus
· Ammagnostus cylindratus Guo and Luo, 1996
· Ammagnostus duibianensis Lu and Lin, 1989
· Ammagnostus histus
· Ammagnostus hunanensis
· Ammagnostus integriceps Öpik, 1967
· Ammagnostus laiwuensis (Lorenz, 1906)
· Ammagnostus mitis Öpik, 1967
· Ammagnostus sinensis Peng, 1987
· Ammagnostus wangcunensis Peng and Robison

Hadragnostus Öpik, 1967

· Hadragnostus las Öpik, 1967 (Type)
· Hadragnostus edax Fortey and Rushton, 1976
· Hadragnostus helixensis Jago and Cooper, 2005
· Hadragnostus modestus (Lochman, 1944)

Kormagnostus Resser, 1938

· Kormagnostus sinplex Resser, 1938 (Type)
· Kormagnostus boltoni Westrop et al., 1996
· Kormagnostus copelandi Westrop et al., 1996
· Kormagnostus flati Pratt, 1992
· Kormagnostus inventus Shergold, 1982
· Kormagnostus minutus (Schrank, 1975)
· Kormagnostus seclusus (Walcott, 1884)

Proagnostus Butts, 1926

· Proagnostus bulbus Butts, 1926 (Type)
· Proagnostus centerensis Resser, 1938
· Proagnostus maryvillensis Resser

They occurred during the late Cambrian period to the Ordovician period.

APC

APC (adenomatosis polyposis coli) is a human gene that is classified as a tumor suppressor gene. Tumor suppressor genes prevent the uncontrolled growth of cells that may result in cancerous tumors. The protein made by the APC gene plays a critical role in several cellular processes that determine whether a cell may develop into a tumor.

Glucagon

Glucagon is an important hormone involved in carbohydrate metabolism. Produced by the pancreas, it is released when the glucose level in the blood is low (hypoglycemia), causing the liver to convert stored glycogen into glucose and release it into the bloodstream. The action of Glucagon is thus opposite to that of insulin, which instructs the body's cells to take in glucose from the blood.

Insulin

Insulin is a hormone that has extensive effects on metabolism and other body functions, such as vascular compliance. Insulin causes cells in the liver, muscle, and fat tissue to take up glucose from the blood, storing it as glycogen in the liver and muscle, and stopping use of fat as an energy source. When Insulin is absent (or low), glucose is not taken up by body cells, and the body begins to use fat as an energy source, for example, by transfer of lipids from adipose tissue to the liver for mobilization as an energy source.

Polyp

A Polyp in zoology is one of two forms found in the phylum Cnidaria, the other being the medusa. Polyp s are approximately cylindrical in shape and elongated at the axis of the body. In solitary Polyp s, the aboral end is attached to the substrate by means of a disc-like holdfast, while in colonies of Polyp s it is connected to other Polyp s, either directly or indirectly.

Juice

JUICE is a widely used non-commercial software package for editing and analysing phytosociological data.

It was developed at the Masaryk University in Brno, Czech Republic in 1998, and is fully described in English manual. It makes use of the previously-developed TURBOVEG software for entering and storing such data) and it offers a quite powerful tool for vegetation data analysis, including:

· creation of synoptic tables
· determination of diagnostic species according to their fidelity
· calculation of Ellenberg indicator values for relevés, various indices of alpha and beta diversity
· classification of relevés using TWINSPAN or cluster analysis
· expert system for vegetation classification based on COCKTAIL method etc. .

Spirogyra

Spirogyra is a genus of filamentous green algae of the order Zygnematales and there are more than 400 species of Spirogyra in the world. Spirogyra measures approximately 10 to 100μm in width and may stretch centimeters long.

Spirogyra is unbranched with cylindrical cells connected end to end in long green filaments.

Digestion

Digestion is the mechanical and chemical breaking down of fat into smaller components, to a form that can be absorbed, for instance, by a blood stream. Digestion is a form of catabolism.

In mammals, food enters the mouth, being chewed by teeth, and broken down by the saliva from the salivary glands.

Maltose

Maltose is a disaccharide formed from two units of glucose joined with an $\alpha(1\rightarrow4)$ linkage. It is the second member of an important biochemical series of glucose chains. The addition of another glucose unit yields maltotriose; further additions will produce dextrins (also called maltodextrins) and eventually starch (glucose polymer.)

Peptide

Peptide s are short polymers formed from the linking, in a defined order, of α-amino acids. The link between one amino acid residue and the next is known as an amide bond or a Peptide bond.

Proteins are poly Peptide molecules .

Genome	In classical genetics, the Genome of a diploid organism including eukarya refers to a full set of chromosomes or genes in a gamete; thereby, a regular somatic cell contains two full sets of Genome s. In haploid organisms, including bacteria, archaea, viruses, and mitochondria, a cell contains only a single set of the Genome usually in a single circular or contiguous linear DNA (or RNA for retroviruses.) In modern molecular biology the Genome of an organism is its hereditary information encoded in DNA (or, for retroviruses, RNA.)
Genome Project	Genome project s are scientific endeavours that ultimately aim to determine the complete genome sequence of an organism (be it an animal, a plant, a fungus, a bacterium, an archaean, a protist or a virus.) The genome sequence for any organism requires the DNA sequences for each of the chromosomes in an organism to be determined. For bacteria, which usually have just one chromosome, a Genome project will aim to map the sequence of that chromosome.
Human	A Human is a member of a species of bipedal primates in the family Hominidae . DNA and fossil evidence indicates that modern Human s originated in east Africa about 200,000 years ago. When compared to other animals and primates, Human s have a highly developed brain, capable of abstract reasoning, language, introspection and problem solving.
Human Genome	The Human genome is the genome of Homo sapiens, which is stored on 23 chromosome pairs. Twenty-two of these are autosomal chromosome pairs, while the remaining pair is sex-determining. The haploid Human genome occupies a total of just over 3 billion DNA base pairs.
Human Genome Project	The Human Genome Project was an international scientific research project with a primary goal to determine the sequence of chemical base pairs which make up DNA and to identify and map the approximately 20,000-25,000 genes of the human genome from both a physical and functional standpoint. The project began in 1990 initially headed by James D. Watson at the U.S. National Institutes of Health. A working draft of the genome was released in 2000 and a complete one in 2003, with further analysis still being published.
Collagen	Collagen is the main protein of connective tissue in animals and the most abundant protein in mammals, making up about 25% to 35% of the whole-body protein content. It is naturally found exclusively in metazoa, including sponges. In muscle tissue it serves as a major component of endomysium.
Decomposers	Decomposers are organisms that consume dead or decaying organisms, and, in doing so, carry out the natural process of decomposition. Like herbivores and predators, Decomposers are heterotrophic, meaning that they use organic substrates to get their energy, carbon and nutrients for growth and development. Decomposers use deceased organisms and non-living organic compounds as their food source.
CD36	CD36 is an integral membrane protein found on the surface of many cell types in vertebrate animals and is also known as FAT, SCARB3, GP88, glycoprotein IV and glycoprotein IIIb CD36 is a member of the class B scavenger receptor family of cell surface proteins. CD36 binds many ligands including collagen, thrombospondin, erythrocytes parasitized with Plasmodium falciparum, oxidized low density lipoprotein, native lipoproteins, oxidized phospholipids, and long-chain fatty acids.

Nutrient

A Nutrient is a chemical that an organism needs to live and grow or a substance used in an organism's metabolism which must be taken in from its environment. Nutrient s are the substances that enrich the body. They build and repair tissues, give heat and energy, and regulate body processes.

Trans fat

Trans fat is the common name for a type of unsaturated fat with trans-isomer fatty acid(s.) Trans fat s may be monounsaturated or polyunsaturated but never saturated.

Unsaturated fat is a fat molecule, containing one or more double bonds between the carbon atoms.

Vitamin C

Vitamin C or L-ascorbic acid is an essential nutrient for humans, a large number of higher primate species, a small number of other mammalian species (notably guinea pigs and bats), a few species of birds, and some fish.

Ascorbate (an ion of ascorbic acid) is required for a range of essential metabolic reactions in all animals and plants. It is made internally by almost all organisms, humans being a notable exception.

Escherichia

Escherichia is a genus of Gram-negative, non-spore forming, facultatively anaerobic, rod-shaped bacteria from the family Enterobacteriaceae. Inhabitants of the gastrointestinal tracts of warm-blooded animals, Escherichia species provide a portion of the microbially-derived vitamin K for their host.

While many Escherichia are harmless commensals, particular strains of some species are human pathogens, and are known as the most common cause of urinary tract infections, significant sources of gastrointestinal disease, ranging from simple diarrhea to dysentery-like conditions, as well as a wide-range of other pathogenic states.

Escherichia coli

Escherichia coli , is a Gram negative bacterium that is commonly found in the lower intestine of warm-blooded organisms . Most E. coli strains are harmless, but some, such as serotype O157:H7, can cause serious food poisoning in humans, and are occasionally responsible for costly product recalls. The harmless strains are part of the normal flora of the gut, and can benefit their hosts by producing vitamin K_2, or by preventing the establishment of pathogenic bacteria within the intestine.

Essential nutrient

An Essential nutrient is a nutrient required for normal body functioning that either cannot be synthesized by the body at all, or cannot be synthesized in amounts adequate for good health (e.g. niacin, choline), and thus must be obtained from a dietary source. Some categories of Essential nutrient s include vitamins, dietary minerals, essential fatty acids, and essential amino acids.

Different species have very different Essential nutrient s.

Micronutrients

Micronutrients are nutrients needed throughout life in small quantities. They are dietary minerals needed by the human body in very small quantities (generally less than 100micrograms/day) as opposed to macrominerals which are required in larger quantities. The Microminerals or trace elements include at least iron, cobalt, chromium, copper, iodine, manganese, selenium, zinc and molybdenum.

Carbohydrates

Carbohydrates or saccharides are the most abundant of the four major classes of biomolecules. They fill numerous roles in living things, such as the storage and transport of energy (eg: starch, glycogen) and structural components (eg: cellulose in plants and chitin.) Additionally, Carbohydrates and their derivatives play major roles in the working process of the immune system, fertilization, pathogenesis, blood clotting, and development.

Class

In biological classification, Class is

· a taxonomic rank. Other well-known ranks are life, domain, kingdom, phylum, order, family, genus, and species, with Class fitting between phylum and order. As for the other well-known ranks, there is the option of an immediately lower rank, indicated by the prefix sub-: subclass .
· a taxonomic unit, a taxon, in that rank. In that case the plural is classes
The composition of each Class is determined by a taxonomist. Often there is no exact agreement, with different taxonomists taking different positions. There are no hard rules that a taxonomist needs to follow in describing a Class, but for well-known animals there is likely to be consensus. For example, dogs are usually assigned to the Class Mammalia; in the phylum Chordata (animals with notochords); in the order Carnivora (mammals that eat meat.)

Acid

An acid is traditionally considered any chemical compound that, when dissolved in water, gives a solution with a hydrogen ion activity greater than in pure water, i.e. a pH less than 7.0. That approximates the modern definition of Johannes Nicolaus Brønsted and Martin Lowry, who independently defined an acid as a compound which donates a hydrogen ion (H^+) to another compound (called a base.) Common examples include acetic acid and sulfuric acid (used in car batteries.)

Essential fatty acid

The actions of the ω-3(Omega-3) and ω-6(Omega-6) essential fatty acids (essential fatty acids) are best characterized by their interactions; they cannot be understood separately.

For introductory details to this topic, including terminology and ω-3 / ω-6 nomenclature, see the main articles at essential fatty acid and Eicosanoid.

Arachidonic acid (AA) is a 20-carbon ω-6 essential fatty acid.

Lipids

Lipids are a broad group of naturally-occurring molecules which includes fats, waxes, sterols, fat-soluble vitamins (such as vitamins A, D, E and K), monoglycerides, diglycerides, phospholipids, and others. The main biological functions of lipids include energy storage, as structural components of cell membranes, and as important signaling molecules.

lipids may be broadly defined as hydrophobic or amphiphilic small molecules; the amphiphilic nature of some lipids allows them to form structures such as vesicles, liposomes, or membranes in an aqueous environment.

Triglycerides

Â·) (more properly known as Â·), TAG or triacylglyceride) is a glyceride in which the glycerol is esterified with three fatty acids. It is the main constituent of vegetable oil and animal fats.

Triglycerides are formed from a single molecule of glycerol, combined with three fatty acids on each of the OH groups, and make up most of fats digested by humans.

Cholesterol

Cholesterol is a lipidic, waxy steroid found in the cell membranes and transported in the blood plasma of all animals. It is an essential component of mammalian cell membranes where it is required to establish proper membrane permeability and fluidity. Cholesterol is the principal sterol synthesized by animals, but small quantities are synthesized in other eukaryotes, such as plants and fungi.

Deamination

Deamination is the removal of an amine group from a molecule.

In the human body, Deamination takes place primarily in the liver, however Glutamate is also deaminated in the kidneys. Deamination is the process by which amino acids are broken down when too much protein has been taken in.

Protein

Protein s are organic compounds made of amino acids arranged in a linear chain. The amino acids in a polymer chain are joined together by the peptide bonds between the carboxyl and amino groups of adjacent amino acid residues. The sequence of amino acids in a protein is defined by the sequence of a gene, which is encoded in the genetic code.

Anemia

Species of the genus Anemia are sometimes called flowering ferns, but this term is more commonly applied to ferns of the genus Osmunda. It is sometimes classified in family Schizaeaceae.

· A. adiantifolia - pineland fern
· A. cicutaria - hemlock fern
· A. coriacia
· A. cuneata
· A. donnell-smithii
· A. hirsuta - hairy flowering fern
· A. hirta - streambank flowering fern
· A. mexicana - Mexican flowering fern
· A. phyllitidis
· A. portoricensis - Puerto Rican flowering fern
· A. underwoodiana
· A. wrightii - Wright's flowering fern .

Minerals

Minerals are required by plants as part of their food, to form their structure. The firmness of straw for example, is due to the presence in it of silica, the principal constituent of sand and flints. Potassa, soda, lime, magnesia, and phosphoric acid are contained in plants, in different proportions.

Antioxidant

An Antioxidant is a molecule capable of slowing or preventing the oxidation of other molecules. Oxidation is a chemical reaction that transfers electrons from a substance to an oxidizing agent. Oxidation reactions can produce free radicals, which start chain reactions that damage cells.

Gene

A Gene is the basic unit of heredity in a living organism. All living things depend on Gene s. Gene s hold the information to build and maintain their cells and pass Gene tic traits to offspring.

Maintenance

Maintenance of an organism is the collection of processes to stay alive, excluding production processes. The Dynamic Energy Budget theory delineates two classes

· Somatic Maintenance. This comprises the turnover of structural mass (mainly proteins), the Maintenance of concentration gradients of metabolites across membranes, activity
· Maturity Maintenance. This comprises the Maintenance of defence systems (such as the immune system), the preparation of the body for reproduction.
The theory assumes that maturity Maintenance costs can be reduced more easily during starvation than somatic Maintenance costs. Under extreme starvation conditions, somatic Maintenance costs are paid from structural mass, which causes shrinking.Some organism manage to switch to the turpor state under starvation conditions, and reduce their Maintenance costs.

Heterochromatin

Heterochromatin is a tightly packed form of DNA. Its major characteristic is that transcription is limited. As such, it is a means to control gene expression, through regulation of the transcription initiation.

Chromatin is found in two varieties: euchromatin and Heterochromatin.

High-density lipoprotein

High-density lipoprotein is one of the five major groups of lipoproteins (chylomicrons, VLDL, IDL, LDL, HDL) which enable lipids like cholesterol and triglycerides to be transported within the water based blood stream. In healthy individuals, about thirty percent of blood cholesterol is carried by HDL .

It is hypothesized that HDL can remove cholesterol from atheroma within arteries and transport it back to the liver for excretion or re-utilization--which is the main reason why HDL-bound cholesterol is sometimes called 'good cholesterol', or HDL-C. A high level of HDL-C seems to protect against cardiovascular diseases, and low HDL cholesterol levels (less than 40 mg/dL) increase the risk for heart disease.

Low-density lipoprotein

Low-density lipoprotein is a type of lipoprotein that transports cholesterol and triglycerides from the liver to peripheral tissues. LDL is one of the five major groups of lipoproteins; these groups include chylomicrons, very Low-density lipoprotein (VLDL), intermediate-density lipoprotein (IDL), Low-density lipoprotein, and high-density lipoprotein (HDL), although some alternative organizational schemes have been proposed. Like all lipoproteins, LDL enables fats and cholesterol to move within the water-based solution of the blood stream.

Unsaturated fat

An Unsaturated fat is a fat or fatty acid in which there are one or more double bonds in the fatty acid chain. A fat molecule is monounsaturated if it contains one double bond, and polyunsaturated if it contains more than one double bond. Where double bonds are formed, hydrogen atoms are eliminated.

Cell

The Cell is the structural and functional unit of all known living organisms. It is the smallest unit of an organism that is classified as living, and is often called the building block of life. Some organisms, such as most bacteria, are unicellular (consist of a single Cell.)

Size

Size has been one of the most interesting aspects of cephalopod science to the general public Extinct taxa are also included.

Decomposers

Decomposers are organisms that consume dead or decaying organisms, and, in doing so, carry out the natural process of decomposition. Like herbivores and predators, Decomposers are heterotrophic, meaning that they use organic substrates to get their energy, carbon and nutrients for growth and development. Decomposers use deceased organisms and non-living organic compounds as their food source.

Horse murders

The Horse murders scandal was a form of insurance fraud in the United States in which expensive horses, many of them show jumpers, were insured against death, accident and then killed to collect the insurance money. It is not known how many horses were murdered between the mid 1970s and the mid-1990s, when a Federal Bureau of Investigation (FBI) investigation brought the horse killings to light, but the number is thought to be well over 50, and may have been as high as 100. In addition, in 1977, the heiress Helen Brach disappeared and was presumed by law enforcement agents to have been murdered by the perpetrators of these crimes, because she threatened to report their criminal activity to authorities; continuing investigations into Brach's death began to uncover the insurance fraud in the 1990s.

Immune system

An immune system is a collection of biological processes within an organism that protects against disease by identifying and killing pathogens and tumour cells. It detects a wide variety of agents, from viruses to parasitic worms, and needs to distinguish them from the organism's own healthy cells and tissues in order to function properly. Detection is complicated as pathogens can evolve rapidly, producing adaptations that avoid the immune system and allow the pathogens to successfully infect their hosts.

APC

APC (adenomatosis polyposis coli) is a human gene that is classified as a tumor suppressor gene. Tumor suppressor genes prevent the uncontrolled growth of cells that may result in cancerous tumors. The protein made by the APC gene plays a critical role in several cellular processes that determine whether a cell may develop into a tumor.

Ephedra

Ephedra, from the plant Ephedra sinica, has been used as an herbal remedy in traditional Chinese medicine for 5,000 years for the treatment of asthma and hay fever, as well as for the common cold. Known in Chinese as ma huang , Ephedra is a stimulant which constricts blood vessels and increases blood pressure and heart rate. Several additional species belonging to the genus Ephedra have traditionally been used for a variety of medicinal purposes, and are a possible candidate for the Soma plant of Indo-Iranian religion.

Spirogyra

Spirogyra is a genus of filamentous green algae of the order Zygnematales and there are more than 400 species of Spirogyra in the world. Spirogyra measures approximately 10 to 100µm in width and may stretch centimeters long.

Spirogyra is unbranched with cylindrical cells connected end to end in long green filaments.

Macrophages

Macrophages are white blood cells within tissues, produced by the division of monocytes. Human Macrophages are about 21 micrometres in diameter. Monocytes and Macrophages are phagocytes, acting in both non-specific defense as well as to help initiate specific defense mechanisms (or adaptive immunity) of vertebrate animals.

Spleen

The Spleen, an organ found in all vertebrate animals, is a major part of the immune system. In humans, the Spleen is located in the abdomen of the body, where it has three primary functions: 1) Removal and destruction of old, aged red blood cells, 2) Synthesis of antibodies in the white pulp, and 3) Removal of antibody-coated bacteria and antibody-coated blood cells from the circulation . It is one of the centers of activity of the reticuloendothelial system, and can be considered analogous to a large lymph node.

Nereis	Nereis is a genus of polychaete worms in the family Nereidae. It comprises many species, most of which are marine, including the sandworm (Nereis virens) and the common clam worm (Nereis succinea.) Nereis possess setae and parapodia for locomotion.
Shigella	Shigella is a genus of Gram-negative, non-spore forming rod-shaped bacteria closely related to Escherichia coli and Salmonella. The causative agent of human shigellosis, Shigella cause disease in primates, but not in other mammals. It is only naturally found in humans and apes.
Shigella dysenteriae	Shigella dysenteriae is a species of the rod-shaped bacterial genus Shigella. Shigella can cause shigellosis (bacillary dysentery.) Shigellae are Gram-negative, non-spore-forming, facultatively anaerobic, non-motile bacteria.
Cell	The Cell is the structural and functional unit of all known living organisms. It is the smallest unit of an organism that is classified as living, and is often called the building block of life. Some organisms, such as most bacteria, are unicellular (consist of a single Cell.)
Histamine	Histamine is a biogenic amine involved in local immune responses as well as regulating physiological function in the gut and acting as a neurotransmitter. Histamine triggers the inflammatory response. As part of an immune response to foreign pathogens, Histamine is produced by basophils and by mast cells found in nearby connective tissues.
Mast	Mast is the edible vegatative or reproductive part produced by woody species of plants, i.e. trees and shrubs, that wildlife species and some domestic animals consume. It comes in two forms. Tree species such as oak, hickory and beech produce a hard Mast - acorns, hickory nuts, and beechnuts.
Mast cell	A Mast cell (or mastocyte) is a resident cell of several types of tissues and contains many granules rich in histamine and heparin. Although best known for their role in allergy and anaphylaxis, Mast cell s play an important protective role as well, being intimately involved in wound healing and defense against pathogens. Mast cell /span> Mast cell s were first described by Paul Ehrlich in his 1878 doctoral thesis on the basis of their unique staining characteristics and large granules.
NADPH	Nicotinamide adenine dinucleotide phosphate ($NADP^+$, in older notation triphosphopyridine nucleotide, TPN) is used in anabolic reactions, such as lipid and nucleic acid synthesis, which require NADPH as a reducing agent. NADPH is the reduced form of $NADP^+$, and $NADP^+$ is the oxidized form of NADPH. NADP+ differs from NAD+ by the presence in NADP+ of an additional phosphate group on the 2' position of the ribose ring that carries the adenine moiety. In chloroplasts, NADP is reduced by ferredoxin-NADP+ reductase in the last step of the electron chain of the light reactions of photosynthesis.
Photosynthesis	Photosynthesis is a process that converts carbon dioxide into organic compounds, especially sugars, using the energy from sunlight. Photosynthesis occurs in plants, algae, and many species of Bacteria, but not in Archaea. Photosynthetic organisms are called photoautotrophs, since it allows them to create their own food.

Acetabularia	Acetabularia is a genus of green algae, specifically of the Polyphysaceae family, Typically found in subtropical waters, Acetabularia is a single-cell organism, but gigantic in size and complex in form, making it an excellent model organism for studying cell biology. In form, the mature Acetabularia resembles the round leaves of a nasturtium, being 0.5 to 10 cm tall and having three anatomical parts: a bottom rhizoid that resembles a set of short roots; a long stalk in the middle; and a top umbrella of branches that may fuse into a cap. The single nucleus of Acetabularia is located in the rhizoid, and allows the cell to regenerate completely if its cap is removed.
Acetylcholine	The chemical compound Acetylcholine is a neurotransmitter in both the peripheral nervous system (PNS) and central nervous system (CNS) in many organisms including humans. Acetylcholine is one of many neurotransmitters in the autonomic nervous system (ANS) and the only neurotransmitter used in the motor division of the somatic nervous system. (Sensory neurons use glutamate and various peptides at their synapses.)
Apoptosis	Apoptosis is the process of programmed cell death that may occur in multicellular organisms. Programmed cell death involves a series of biochemical events leading to a characteristic cell morphology and death, in more specific terms, a series of biochemical events that lead to a variety of morphological changes, including blebbing, changes to the cell membrane such as loss of membrane asymmetry and attachment, cell shrinkage, nuclear fragmentation, chromatin condensation, and chromosomal DNA fragmentation (1-4.)
Plasma cells	Plasma cells plasmocytes, and effector B cells, are white blood cells that produce large volumes of antibodies. They are transported by the blood plasma and the lymphatic system. Like all blood cells, Plasma cells ultimately originate in the bone marrow; however, these cells leave the bone marrow as B cells, before terminal differentiation into Plasma cells, which usually happens in the spleen or lymph nodes.
Programmed cell death	Programmed cell-death (or programmed cell death) is death of a cell in any form, mediated by an intracellular program. In contrast to necrosis, which is a form of cell-death that results from acute tissue injury and provokes an inflammatory response, programmed cell death is carried out in a regulated process which generally confers advantage during an organism's life-cycle. programmed cell death serves fundamental functions during both plant and metazoa (multicellular animals) tissue development.
Receptor	In biochemistry, a Receptor is a protein molecule, embedded in either the plasma membrane or cytoplasm of a cell, to which a mobile signaling (or 'signal') molecule may attach. A molecule which binds to a Receptor is called a 'ligand,' and may be a peptide (such as a neurotransmitter), a hormone, a pharmaceutical drug, or a toxin, and when such binding occurs, the Receptor undergoes a conformational change which ordinarily initiates a cellular response. However, some ligands merely block Receptor s without inducing any response (e.g. antagonists.)
Blood type	A Blood type is a classification of blood based on the presence or absence of inherited antigenic substances on the surface of red blood cells These antigens may be proteins, carbohydrates, glycoproteins depending on the blood group system, and some of these antigens are also present on the surface of other types of cells of various tissues. Several of these red blood cell surface antigens, that stem from one allele, collectively form a blood group system.
Cystic fibrosis	Cystic fibrosis is a genetic disorder known to be an inherited disease of the secretory glands, including the glands that make mucus and sweat.

	The hallmarks of Cystic fibrosis are salty tasting skin, normal appetite but poor growth and poor weight gain, excess mucus production, and coughing/shortness of breath. Males can be infertile due to the condition congenital bilateral absence of the vas deferens.
Helicobacter	Helicobacter is a genus of Gram-negative bacteria possessing a characteristic helix shape. They were initially considered to be members of the Campylobacter genus, but since 1989 they have been grouped in their own genus. Some species have been found living in the lining of the upper gastrointestinal tract, as well as the liver of mammals and some birds..
Helicobacter pylori	Helicobacter pylori is a Gram-negative, microaerophilic bacterium that inhabits various areas of the stomach and duodenum. It causes a chronic low-level inflammation of the stomach lining and is strongly linked to the development of duodenal and gastric ulcers and stomach cancer. Over 80% of individuals infected with the bacterium are asymptomatic.
Gamma	Gamma is a small neotropical, primarily northern andean genus of potter wasps currently containing 6 recognized species.
Vaccine	A vaccine is a biological preparation that improves immunity to a particular disease. A vaccine typically contains a small amount of an agent that resembles a microorganism. The agent stimulates the body's immune system to recognize the agent as foreign, destroy it, and 'remember' it, so that the immune system can more easily recognize and destroy any of these microorganisms that it later encounters.
Alagille syndrome	Alagille syndrome is a genetic disorder that affects the liver, heart, and other systems of the body. Problems associated with the disorder generally become evident in infancy or early childhood. The disorder is inherited in an autosomal dominant pattern, and the estimated prevalence of Alagille syndrome is 1 in every 100,000 live births.
Allergen	An allergen is a nonparasitic antigen capable of stimulating a type-I hypersensitivity reaction in atopic individuals. Most humans mount significant Immunoglobulin E responses only as a defense against parasitic infections. However, some individuals mount an IgE response against common environmental antigens.
Gene	A Gene is the basic unit of heredity in a living organism. All living things depend on Gene s. Gene s hold the information to build and maintain their cells and pass Gene tic traits to offspring.
Pneumonia	Pneumonia is an illness which can result from a variety of causes, including infection with bacteria, viruses, fungi, or parasites. Pneumonia can occur in any animal with lungs, including mammals, birds, and reptiles. Symptoms associated with Pneumonia include fever, fast or difficult breathing, nasal discharge, and decreased activity.

Nereis	Nereis is a genus of polychaete worms in the family Nereidae. It comprises many species, most of which are marine, including the sandworm (Nereis virens) and the common clam worm (Nereis succinea.) Nereis possess setae and parapodia for locomotion.
Human	A Human is a member of a species of bipedal primates in the family Hominidae . DNA and fossil evidence indicates that modern Human s originated in east Africa about 200,000 years ago. When compared to other animals and primates, Human s have a highly developed brain, capable of abstract reasoning, language, introspection and problem solving.
Axon	An Axon or nerve fiber is a long, slender projection of a nerve cell that conducts electrical impulses away from the neuron's cell body or soma. An Axon is one of two types of protoplasmic protrusions that extrude from the cell body of a neuron, the other type being dendrites. Axon s are distinguished from dendrites by several features, including shape (dendrites often taper while Axon s usually maintain a constant radius), length (dendrites are restricted to a small region around the cell body while Axon s can be much longer), and function (dendrites usually receive signals while Axon s usually transmit them.)
Cell	The Cell is the structural and functional unit of all known living organisms. It is the smallest unit of an organism that is classified as living, and is often called the building block of life. Some organisms, such as most bacteria, are unicellular (consist of a single Cell.)
Decomposers	Decomposers are organisms that consume dead or decaying organisms, and, in doing so, carry out the natural process of decomposition. Like herbivores and predators, Decomposers are heterotrophic, meaning that they use organic substrates to get their energy, carbon and nutrients for growth and development. Decomposers use deceased organisms and non-living organic compounds as their food source.
Dendrites	Dendrites are the branched projections of a neuron that act to conduct the electrochemical stimulation received from other neural cells to the cell body of the neuron from which the Dendrites project. Electrical stimulation is transmitted onto Dendrites by upstream neurons via synapses which are located at various points throughout the dendritic arbor. Dendrites play a critical role in integrating these synaptic inputs and in determining the extent to which action potentials are produced by the neuron.
Interneuron	An Interneuron is a multipolar neuron which connects afferent neurons and efferent neurons in neural pathways. Like motor neurons, Interneuron cell bodies are always located in the central nervous system When contrasted with the peripheral nervous system, the neurons of the central nervous system, including the brain, are all Interneuron s.
Myelin	Myelin is a dielectric material that forms a layer, the Myelin sheath. Usually, Myelin surrounds only the axon of a neuron. It is essential for proper functioning of the nervous system.
Nodes of Ranvier	Nodes of Ranvier are the gaps (approximately 1 micrometer in length) formed between the myelin sheaths generated by different cells. A myelin sheath is a many-layered coating, largely composed of a fatty substance called myelin, that wraps around the axon of a neuron and very efficiently insulates it. At Nodes of Ranvier, the axonal membrane is uninsulated and therefore capable of generating electrical activity.

Salmonella

Salmonella is a genus of rod-shaped, Gram-negative, non-spore forming, predominantly motile enterobacteria with diameters around 0.7 to 1.5 Åμm, lengths from 2 to 5 Åμm, and flagella which project in all directions (i.e. peritrichous.) They are chemoorganotrophs, obtaining their energy from oxidation and reduction reactions using organic sources and are facultative anaerobes; most species produce hydrogen sulfide, which can readily be detected by growing them on media containing ferrous sulfate, such as TSI. Most isolates exist in two phases; phase I is the motile phase and phase II the non-motile phase. Cultures that are non-motile upon primary culture may be swithched to the motile phase using a Craigie tube.

Acetabularia

Acetabularia is a genus of green algae, specifically of the Polyphysaceae family, Typically found in subtropical waters, Acetabularia is a single-cell organism, but gigantic in size and complex in form, making it an excellent model organism for studying cell biology. In form, the mature Acetabularia resembles the round leaves of a nasturtium, being 0.5 to 10 cm tall and having three anatomical parts: a bottom rhizoid that resembles a set of short roots; a long stalk in the middle; and a top umbrella of branches that may fuse into a cap. The single nucleus of Acetabularia is located in the rhizoid, and allows the cell to regenerate completely if its cap is removed.

Drosophila

Drosophila has long been a favorite model system for geneticists and developmental biologists studying embryogenesis. The small size, short generation time, and large brood size makes it ideal for genetic studies. Transparent embryos facilitate developmental studies.

Acetylcholine

The chemical compound Acetylcholine is a neurotransmitter in both the peripheral nervous system (PNS) and central nervous system (CNS) in many organisms including humans. Acetylcholine is one of many neurotransmitters in the autonomic nervous system (ANS) and the only neurotransmitter used in the motor division of the somatic nervous system. (Sensory neurons use glutamate and various peptides at their synapses.)

Acetylcholinesterase

Acetylcholinesterase is an enzyme that degrades the neurotransmitter acetylcholine, producing choline and an acetate group. It is mainly found at neuromuscular junctions and cholinergic synapses in the central nervous system, where its activity serves to terminate synaptic transmission. AChE has a very high catalytic activity -- each molecule of AChE degrades about 5000 molecules of acetylcholine per second.

Depressants

Depressants are psychoactive drugs which temporarily diminish the function or activity of a specific part of the body or mind. Examples of these kinds of effects may include anxiolysis, sedation, and hypotension. Due to their effects typically having a 'down' quality to them, Depressants are also occasionally referred to as 'downers'.

Neurotransmitters

Neurotransmitters are endogenous chemicals which relay, amplify, and modulate signals between a neuron and another cell. Neurotransmitters are packaged into synaptic vesicles that cluster beneath the membrane on the presynaptic side of a synapse, and are released into the synaptic cleft, where they bind to receptors in the membrane on the postsynaptic side of the synapse. Release of Neurotransmitters usually follows arrival of an action potential at the synapse, but may follow graded electrical potentials.

Norepinephrine

Noradrenaline (BAN) or Norepinephrine is a catecholamine with dual roles as a hormone and a neurotransmitter.

As a stress hormone, Norepinephrine affects parts of the brain where attention and responding actions are controlled. Along with epinephrine, Norepinephrine also underlies the fight-or-flight response, directly increasing heart rate, triggering the release of glucose from energy stores, and increasing blood flow to skeletal muscle.

Stimulants	Stimulants also sometimes called psycho Stimulants , are psychoactive drugs which induce temporary improvements in either mental or physical function or both. Examples of these kinds of effects may include enhanced alertness, wakefulness, and locomotion, among others. Due to their effects typically having an 'up' quality to them, Stimulants are also occasionally referred to as 'uppers'.
Crassulacean acid metabolism	Crassulacean acid metabolism is an elaborate carbon fixation pathway in some plants. These plants fix carbon dioxide during the night, storing it as the four carbon acid malate. The CO_2 is released during the day, where it is concentrated around the enzyme RuBisCO, increasing the efficiency of photosynthesis.
Cannabis	Cannabis is a genus of flowering plants that includes three putative species, Cannabis sativa L., Cannabis indica Lam., and Cannabis ruderalis Janisch. These three taxa are indigenous to Central Asia, South Asia, and surrounding regions. Cannabis has long been used for fibre (hemp), for medicinal purposes, and as a recreational drug.
Cocaine	The first synthesis and elucidation of the cocaine molecule was by Richard Willstätter in 1898. Willstätter's synthesis derived cocaine from tropinone. Since then, Robert Robinson and Edward Leete have made significant contributions to the mechanism of the synthesis.
Marijuana	The term Marijuana for cannabis, in English-language usage, dates to the fears of Spanish immigrants carrying the drug A report by the U.S. Federal Bureau of Narcotics had found, 'This abuse of the drug is noted among the Latin-American or Spanish-speaking population. The sale of cannabis cigarettes occurs to a considerable degree in States along the Mexican border and in cities of the Southwest and West, as well as in New York City and, in fact, wherever there are settlements of Latin Americans.' According to Jack Herer, William Randolph Hearst, 'through pervasive and repetitive use, pounded the obscure Mexican slang word 'Marijuana' into the English-speaking American consciousness. Meanwhile, the word 'hemp' was discarded and 'cannabis,' the scientific term, was ignored and buried.
Cerebrum	The Cerebrum or telencephalon, together with the diencephalon, constitute the forebrain. It is the most anterior or, especially in humans, most superior region of the vertebrate central nervous system. 'Telencephalon' refers to the embryonic structure, from which the mature 'Cerebrum' develops.
Cortex	In botany, the Cortex is the outer of the stem or root of a plant, bounded on the outside by the epidermis and on the inside by the endodermis. It is composed mostly of undifferentiated cells, usually large thin-walled parenchyma cells of the ground tissue system. The outer cortical cells often acquire irregularly thickened cell walls, and are called collenchyma cells.
Ascaris	Ascaris is a genus of parasitic nematode worms known as the giant intestinal roundworms. One species, A. suum, typically infects pigs, while another, A. lumbricoides, affects human populations, typically in sub-tropical and tropical areas with poor sanitation. A. lumbricoides is the largest intestinal roundworm and is the most common helminth infection of humans worldwide, an infection known as ascariasis.
Ascaris lumbricoides	Ascaris lumbricoides is the member of the Ascaris family responsible for the disease ascariasis. It can reach a length of up to 35 cm. Ascaris lumbricoides, or 'roundworm', infections in humans occur when an ingested infective egg releases a larval worm that penetrates the wall of the duodenum and enters the bloodstream.

Diencephalon

The Diencephalon is the region of the brain that includes the thalamus, hypothalamus, epithalamus, prethalamus or subthalamus and pretectum. The Diencephalon is located at the midline of the brain, above the mesencephalon of the brain stem. The Diencephalon contains the zona limitans intrathalamica as morphological boundary and signalling center between the prethalamus and the thalamus.

Melatonin

Melatonin , also known chemically as N-acetyl-5-methoxytryptamine, is a naturally occurring hormone found in animals and in some other living organisms, including algae. Circulating levels vary in a daily cycle, and Melatonin is important in the regulation of the circadian rhythms of several biological functions. Many biological effects of Melatonin are produced through activation of Melatonin receptors, while others are due to its role as a pervasive and powerful antioxidant with a particular role in the protection of nuclear and mitochondrial DNA.

Products containing Melatonin have been available as a dietary supplement in the United States since 1993.

Stem

A stem is one of two main structural axes of a vascular plant. The stem is normally divided into nodes and internodes, the nodes hold buds which grow into one or more leaves, inflorescence (flowers), cones or other stems etc. The internodes act as spaces that distance one node from another.

Shigella

Shigella is a genus of Gram-negative, non-spore forming rod-shaped bacteria closely related to Escherichia coli and Salmonella. The causative agent of human shigellosis, Shigella cause disease in primates, but not in other mammals. It is only naturally found in humans and apes.

Shigella dysenteriae

Shigella dysenteriae is a species of the rod-shaped bacterial genus Shigella. Shigella can cause shigellosis (bacillary dysentery.) Shigellae are Gram-negative, non-spore-forming, facultatively anaerobic, non-motile bacteria.

Horse murders

The Horse murders scandal was a form of insurance fraud in the United States in which expensive horses, many of them show jumpers, were insured against death, accident and then killed to collect the insurance money. It is not known how many horses were murdered between the mid 1970s and the mid-1990s, when a Federal Bureau of Investigation (FBI) investigation brought the horse killings to light, but the number is thought to be well over 50, and may have been as high as 100. In addition, in 1977, the heiress Helen Brach disappeared and was presumed by law enforcement agents to have been murdered by the perpetrators of these crimes, because she threatened to report their criminal activity to authorities; continuing investigations into Brach's death began to uncover the insurance fraud in the 1990s.

DNA

Deoxyribonucleic acid (DNA) is a nucleic acid that contains the genetic instructions used in the development and functioning of all known living organisms and some viruses. The main role of DNA molecules is the long-term storage of information. DNA is often compared to a set of blueprints or a recipe, or a code, since it contains the instructions needed to construct other components of cells, such as proteins and RNA molecules.

Reflex

A Reflex action is an involuntary and nearly instantaneous movement in response to a stimulus. In most contexts, in particular those involving humans, Reflex actions are mediated via the Reflex arc; this is not always true in other animals, nor does it apply to casual uses of the term 'Reflex'.

For a Reflex, reaction time or latency is the time from the onset of a stimulus until the organism responds.

Root	In vascular plants, the Root is the organ of a plant that typically lies below the surface of the soil. This is not always the case, however, since a Root can also be aerial (growing above the ground) or aerating (growing up above the ground or especially above water.) Furthermore, a stem normally occurring below ground is not exceptional either
Stimulus	In physiology, a Stimulus is a detectable change in the internal or external environment. The ability of an organism or organ to respond to external stimuli is called sensitivity. When a Stimulus is applied to a sensory receptor, it elicits or influences a reflex via Stimulus transduction.
Technology	Technology is a broad concept that deals with an animal species' ethology or behavior of usage and of knowledge of tools and crafts, and how it affects the animal species' ability to control and adapt to its environment. Technology is a term with origins in the Greek 'technologia', 'τεχνολογΐ¯α' -- 'techne', 'τΐχνη' and 'logia', 'λογΐ¯α' ('saying'.) However, a strict definition is elusive; 'Technology' can refer to material objects of use to humanity, such as machines, hardware or utensils, but can also encompass broader themes, including systems, methods of organization, and techniques.
Division	Division, in horticulture and gardening, is a method of asexual plant propagation, where the plant (usually an herbaceous perennial) is broken up into two or more parts. Both the root and crown of each part is kept intact. The technique is of ancient origin, and has long been used to propagate bulbs such as garlic and saffron.
Endocrine system	The Endocrine system is a system of glands that involve the release of extracellular signaling molecules known as hormones. The Endocrine system is instrumental in regulating metabolism, growth, development and puberty, and tissue function and also plays a part in determining mood. The field of study that deals with disorders of endocrine glands is endocrinology, a branch of the wider field of internal medicine.
Hormone	Hormone s are chemicals released by cells that affect cells in other parts of the body. Only a small amount of Hormone is required to alter cell metabolism. It is essentially a chemical messenger that transports a signal from one cell to another.
Steroid	A Steroid is a terpenoid lipid characterized by its sterane or Steroid nucleus: a carbon skeleton with four fused rings, generally arranged in a 6-6-6-5 fashion. Steroid s vary by the functional groups attached to these rings and the oxidation state of the rings. Hundreds of distinct Steroid s are found in plants, animals, and fungi.
Adrenocorticotropic hormone	Adrenocorticotropic hormone is a polypeptide tropic hormone produced and secreted by the anterior pituitary gland. It is an important component of the hypothalamic-pituitary-adrenal axis and is often produced in response to biological stress (along with corticotropin-releasing hormone from the hypothalamus.) Its principal effects are increased production of androgens and, as its name suggests, cortisol from the adrenal cortex.
Follicle-stimulating hormone	Follicle-stimulating hormone is a hormone synthesized and secreted by gonadotropes in the anterior pituitary gland. FSH regulates the development, growth, pubertal maturation, and reproductive processes of the human body. FSH and Luteinizing hormone (LH) act synergistically in reproduction.
Luteinizing hormone	Luteinizing hormone is a hormone produced by the anterior pituitary gland. · In the female, an acute rise of Luteinizing hormone - the Luteinizing hormone surge - triggers ovulation.

· In the male, where Luteinizing hormone had also been called Interstitial Cell Stimulating Hormone , it stimulates Leydig cell production of testosterone.

Luteinizing hormone is a glycoprotein. Each monomeric unit is a sugar-like protein molecule; two of these make the full, functional protein.

Peptide

Peptide s are short polymers formed from the linking, in a defined order, of α-amino acids. The link between one amino acid residue and the next is known as an amide bond or a Peptide bond.

Proteins are poly Peptide molecules .

Peptide hormones

Peptide hormones are a class of peptides that are secreted into the blood stream and have endocrine functions in living animals. Peptide hormones are increasingly being identified in plants with important roles in cell-to-cell communication and plant defence. Plant Peptide hormones predominantly act as ligands to membrane-bound receptor kinases.

Prolactin

Prolactin or Luteotropic hormone (LTH) is a peptide hormone primarily associated with lactation. In breastfeeding, the act of an infant suckling the nipple stimulates the production of Prolactin, which fills the breast with milk via a process called lactogenesis, in preparation for the next feed. Oxytocin, another hormone, is also released, which triggers milk let-down.

Thyroid-stimulating hormone

Thyroid-stimulating hormone is a peptide hormone synthesized and secreted by thyrotrope cells in the anterior pituitary gland which regulates the endocrine function of the thyroid gland.

TSH stimulates the thyroid gland to secrete the hormones thyroxine (T_4) and triiodothyronine ($T_{3.}$) TSH production is controlled by a Thyrotropin Releasing Hormone, (TRH), which is manufactured in the hypothalamus and transported to the anterior pituitary gland via the superior hypophyseal artery, where it increases TSH production and release.

Oxytocin

Oxytocin is a mammalian hormone that also acts as a neurotransmitter in the brain.

It is best known for its roles in female reproduction: it is released in large amounts after distension of the cervix and vagina during labor, and after stimulation of the nipples, facilitating birth and breastfeeding, respectively. Recent studies have begun to investigate Oxytocin's role in various behaviors, including orgasm, social recognition, pair bonding, anxiety, trust, love, and maternal behaviors.

Calcitonin

Calcitonin is a 32-amino acid linear polypeptide hormone that is produced in humans primarily by the parafollicular cells of the thyroid, and in many other animals in the ultimobranchial body. It acts to reduce blood calcium, opposing the effects of parathyroid hormone It has been found in fish, reptiles, birds, and mammals.

Iodine

Iodine , is a chemical element that has the symbol I and atomic number 53. Naturally-occurring Iodine is a single isotope with 74 neutrons.

Chemically, Iodine is the second least reactive of the halogens, and the second most electropositive halogen; trailing behind astatine in both of these categories.

Parathyroid hormone	Parathyroid hormone is secreted by the parathyroid glands as a polypeptide containing 84 amino acids. It acts to increase the concentration of calcium (Ca^{2+}) in the blood, whereas calcitonin (a hormone produced by the parafollicular cells (C cells) of the thyroid gland) acts to decrease calcium concentration. PTH acts to increase the concentration of calcium in the blood by acting upon Parathyroid hormone receptor in three parts of the body: PTH half-life is approximately 4 minutes.
Alagille syndrome	Alagille syndrome is a genetic disorder that affects the liver, heart, and other systems of the body. Problems associated with the disorder generally become evident in infancy or early childhood. The disorder is inherited in an autosomal dominant pattern, and the estimated prevalence of Alagille syndrome is 1 in every 100,000 live births.
Coral	Coral s are marine organisms from the class Anthozoa and exist as small sea anemone-like polyps, typically in colonies of many identical individuals. The group includes the important reef builders that are found in tropical oceans, which secrete calcium carbonate to form a hard skeleton. A Coral 'head', commonly perceived to be a single organism, is formed from myriads of individual but genetically identical polyps, each polyp only a few millimeters in diameter.
Globular protein	Globular protein s comprising 'globe'-like proteins that are more or less soluble in aqueous solutions (where they form colloidal solutions.) This main characteristic helps distinguishing them from fibrous proteins (the other class), which are practically insoluble. The term globin can refer more specifically to proteins including the globin fold.
Glucagon	Glucagon is an important hormone involved in carbohydrate metabolism. Produced by the pancreas, it is released when the glucose level in the blood is low (hypoglycemia), causing the liver to convert stored glycogen into glucose and release it into the bloodstream. The action of Glucagon is thus opposite to that of insulin, which instructs the body's cells to take in glucose from the blood.
Insulin	Insulin is a hormone that has extensive effects on metabolism and other body functions, such as vascular compliance. Insulin causes cells in the liver, muscle, and fat tissue to take up glucose from the blood, storing it as glycogen in the liver and muscle, and stopping use of fat as an energy source. When Insulin is absent (or low), glucose is not taken up by body cells, and the body begins to use fat as an energy source, for example, by transfer of lipids from adipose tissue to the liver for mobilization as an energy source.

Calcium

Calcium is a common signaling mechanism, as once it enters the cytoplasm it exerts allosteric regulatory affects on many enzymes and proteins. Calcium can act in signal transduction after influx resulting from activation of ion channels or as a second messenger caused by indirect signal transduction pathways such as G protein-coupled receptors.

Movement of Calcium ions from the extracellular compartment to the intracellular compartment alters membrane depolarisation.

Stimulus

In physiology, a Stimulus is a detectable change in the internal or external environment. The ability of an organism or organ to respond to external stimuli is called sensitivity. When a Stimulus is applied to a sensory receptor, it elicits or influences a reflex via Stimulus transduction.

Salmonella

Salmonella is a genus of rod-shaped, Gram-negative, non-spore forming, predominantly motile enterobacteria with diameters around 0.7 to 1.5 Âµm, lengths from 2 to 5 Âµm, and flagella which project in all directions (i.e. peritrichous.) They are chemoorganotrophs, obtaining their energy from oxidation and reduction reactions using organic sources and are facultative anaerobes; most species produce hydrogen sulfide, which can readily be detected by growing them on media containing ferrous sulfate, such as TSI. Most isolates exist in two phases; phase I is the motile phase and phase II the non-motile phase. Cultures that are non-motile upon primary culture may be swithched to the motile phase using a Craigie tube.

Shigella

Shigella is a genus of Gram-negative, non-spore forming rod-shaped bacteria closely related to Escherichia coli and Salmonella. The causative agent of human shigellosis, Shigella cause disease in primates, but not in other mammals. It is only naturally found in humans and apes.

Shigella dysenteriae

Shigella dysenteriae is a species of the rod-shaped bacterial genus Shigella. Shigella can cause shigellosis (bacillary dysentery.) Shigellae are Gram-negative, non-spore-forming, facultatively anaerobic, non-motile bacteria.

Bud

In botany, a Bud is an undeveloped or embryonic shoot and normally occurs in the axil of a leaf or at the tip of the stem. Once formed, a Bud may remain for some time in a dormant condition, or it may form a shoot immediately.

The Bud s of many woody plants, especially in temperate or cold climates, are protected by a covering of modified leaves called scales which tightly enclose the more delicate parts of the Bud

Chemical

A chemical substance is a material with a specific chemical composition.

A common example of a chemical substance is pure water; it has the same properties and the same ratio of hydrogen to oxygen whether it is isolated from a river or made in a laboratory. Some typical chemical substances are diamond, gold, salt (sodium chloride) and sugar (sucrose.)

Sense

Sense s are the physiological methods of perception. The Sense s and their operation, classification, and theory are overlapping topics studied by a variety of fields, most notably neuroscience, cognitive psychology (or cognitive science), and philosophy of perception. The nervous system has a specific sensory system, or organ, dedicated to each Sense

Citric acid

Citric acid is a weak organic acid, and it is a natural preservative and is also used to add an acidic taste to foods and soft drinks. In biochemistry, it is important as an intermediate in the Citric acid cycle and therefore occurs in the metabolism of virtually all living things. It can also be used as an environmentally benign cleaning agent.

Spirogyra

Spirogyra is a genus of filamentous green algae of the order Zygnematales and there are more than 400 species of Spirogyra in the world. Spirogyra measures approximately 10 to 100µm in width and may stretch centimeters long.

Spirogyra is unbranched with cylindrical cells connected end to end in long green filaments.

Cell

The Cell is the structural and functional unit of all known living organisms. It is the smallest unit of an organism that is classified as living, and is often called the building block of life. Some organisms, such as most bacteria, are unicellular (consist of a single Cell.)

Horse murders

The Horse murders scandal was a form of insurance fraud in the United States in which expensive horses, many of them show jumpers, were insured against death, accident and then killed to collect the insurance money. It is not known how many horses were murdered between the mid 1970s and the mid-1990s, when a Federal Bureau of Investigation (FBI) investigation brought the horse killings to light, but the number is thought to be well over 50, and may have been as high as 100. In addition, in 1977, the heiress Helen Brach disappeared and was presumed by law enforcement agents to have been murdered by the perpetrators of these crimes, because she threatened to report their criminal activity to authorities; continuing investigations into Brach's death began to uncover the insurance fraud in the 1990s.

Ear

The Ear is the organ that detects sound. The vertebrate Ear shows a common biology from fish to humans, with variations in structure according to order and species. It not only acts as a receiver for sound, but plays a major role in the sense of balance and body position.

Membrane

A Membrane is a layer of material which serves as a selective barrier between two phases and remains impermeable to specific particles, molecules, or substances when exposed to the action of a driving force. Some components are allowed passage by the Membrane into a permeate stream, whereas others are retained by it and accumulate in the retentate stream.

Membrane s can be of various thickness, with homogeneous or heterogeneous structure.

ATP synthase

An ATP synthase is a general term for an enzyme that can synthesize adenosine triphosphate (ATP) from adenosine diphosphate (ADP) and inorganic phosphate by using some form of energy. This energy is often in the form of protons moving down an electrochemical gradient, such as from the lumen into the stroma of chloroplasts or from the inter-membrane space into the matrix in mitochondria. The overall reaction sequence is:

$$ADP + P_i \rightarrow ATP$$

These enzymes are of crucial importance in almost all organisms, because ATP is the common 'energy currency' of cells.

Nereis

Nereis is a genus of polychaete worms in the family Nereidae. It comprises many species, most of which are marine, including the sandworm (Nereis virens) and the common clam worm (Nereis succinea.) Nereis possess setae and parapodia for locomotion.

RNA

Ribonucleic acid (RNA) is a biologically important type of molecule that consists of a long chain of nucleotide units. Each nucleotide consists of a nitrogenous base, a ribose sugar, and a phosphate. RNA is very similar to DNA, but differs in a few important structural details: in the cell, RNA is usually single-stranded, while DNA is usually double-stranded; RNA nucleotides contain ribose while DNA contains deoxyribose (a type of ribose that lacks one oxygen atom); and RNA has the base uracil rather than thymine that is present in DNA.

RNA is transcribed from DNA by enzymes called RNA polymerases and is generally further processed by other enzymes.

Lateral line

In aquatic organisms (chiefly fish), the Lateral line is a sense organ used to detect movement and vibration in the surrounding water. Lateral line s are usually visible as faint lines running lengthwise down each side, from the vicinity of the gill covers to the base of the tail. Sometimes parts of the lateral organ are modified into electroreceptors, which are organs used to detect electrical impulses.

Receptor

In biochemistry, a Receptor is a protein molecule, embedded in either the plasma membrane or cytoplasm of a cell, to which a mobile signaling (or 'signal') molecule may attach. A molecule which binds to a Receptor is called a 'ligand,' and may be a peptide (such as a neurotransmitter), a hormone, a pharmaceutical drug, or a toxin, and when such binding occurs, the Receptor undergoes a conformational change which ordinarily initiates a cellular response. However, some ligands merely block Receptor s without inducing any response (e.g. antagonists.)

Calvin cycle

The Calvin cycle is a series of biochemical reactions that take place in the stroma of chloroplasts in photosynthetic organisms. It was discovered by Melvin Calvin, James Bassham and Andrew Benson at the University of California, Berkeley . It is one of the light-independent reactions or dark reactions.

Punnett square

The Punnett square is a diagram that is used to predict the outcome of a particular cross or breeding experiment. It is named after Reginald C. Punnett, who devised the approach, and is used by biologists to determine the probability of an offspring having a particular genotype. The Punnett square is a summary of every possible combination of one maternal allele with one paternal allele for each gene being studied in the cross.

Chromosome

A Chromosome is an organized structure of DNA and protein that is found in cells. It is a single piece of coiled DNA containing many genes, regulatory elements and other nucleotide sequences. Chromosome s also contain DNA-bound proteins, which serve to package the DNA and control its functions.

Gene

A Gene is the basic unit of heredity in a living organism. All living things depend on Gene s. Gene s hold the information to build and maintain their cells and pass Gene tic traits to offspring.

Vision

Vision is the most important sense for birds, since good eyesight is essential for safe flight, and this group has a number of adaptations which give visual acuity superior to that of other vertebrate groups; a pigeon has been described as 'two eyes with wings'. The avian eye resembles that of a reptile, but has a better-positioned lens, a feature shared with mammals. Birds have the largest eyes relative to their size within the animal kingdom, and movement is consequently limited within the eye's bony socket.

Acid

An acid is traditionally considered any chemical compound that, when dissolved in water, gives a solution with a hydrogen ion activity greater than in pure water, i.e. a pH less than 7.0. That approximates the modern definition of Johannes Nicolaus Brønsted and Martin Lowry, who independently defined an acid as a compound which donates a hydrogen ion (H^+) to another compound (called a base.) Common examples include acetic acid and sulfuric acid (used in car batteries.)

Cone

A cone is an organ on plants in the division Pinophyta (conifers) that contains the reproductive structures. The familiar woody cone is the female cone, which produces seeds. The male cones, which produce pollen, are usually herbaceous and much less conspicuous even at full maturity.

Spot

Spots are a method of smoking cannabis . In this method, small pieces of cannabis are rolled to form the 'Spot'. Generally, the tips of two knife blades are heated, the Spot is compressed between the two blades, and the subsequent smoke is inhaled.

Vitamin A

Vitamin A, a bi-polar molecule formed with bi-polar covalent bonds between carbon and hydrogen, is linked to a family of similarly shaped molecules, the retinoids, which complete the remainder of the vitamin sequence. Its important part is the retinyl group, which can be found in several forms. In foods of animal origin, the major form of vitamin A is an ester, primarily retinyl palmitate, which is converted to an alcohol (retinol) in the small intestine.

Endoskeleton

An Endoskeleton is an internal support structure of an animal, composed of mineralized tissue. In three phyla and one subclass of animals, Endoskeleton s of various complexity are found: Chordata, Echinodermata, Porifera, and Coleoidea. An Endoskeleton may function purely for support (as in the case of sponges), but often serves as an attachment site for muscle and a mechanism for transmitting muscular forces.

Exoskeleton

An Exoskeleton is an external skeleton that supports and protects an animal's body, in contrast to the internal endoskeleton of, for example, a human. Some animals, such as the tortoise, have both an endoskeleton and an Exoskeleton In popular usage, many of the larger kinds of Exoskeleton s are known as 'shells'.

APC

APC (adenomatosis polyposis coli) is a human gene that is classified as a tumor suppressor gene. Tumor suppressor genes prevent the uncontrolled growth of cells that may result in cancerous tumors. The protein made by the APC gene plays a critical role in several cellular processes that determine whether a cell may develop into a tumor.

Genome

In classical genetics, the Genome of a diploid organism including eukarya refers to a full set of chromosomes or genes in a gamete; thereby, a regular somatic cell contains two full sets of Genome s. In haploid organisms, including bacteria, archaea, viruses, and mitochondria, a cell contains only a single set of the Genome usually in a single circular or contiguous linear DNA (or RNA for retroviruses.) In modern molecular biology the Genome of an organism is its hereditary information encoded in DNA (or, for retroviruses, RNA.)

Genome Project

Genome project s are scientific endeavours that ultimately aim to determine the complete genome sequence of an organism (be it an animal, a plant, a fungus, a bacterium, an archaean, a protist or a virus.) The genome sequence for any organism requires the DNA sequences for each of the chromosomes in an organism to be determined. For bacteria, which usually have just one chromosome, a Genome project will aim to map the sequence of that chromosome.

Human

A Human is a member of a species of bipedal primates in the family Hominidae . DNA and fossil evidence indicates that modern Human s originated in east Africa about 200,000 years ago. When compared to other animals and primates, Human s have a highly developed brain, capable of abstract reasoning, language, introspection and problem solving.

Human Genome	The Human genome is the genome of Homo sapiens, which is stored on 23 chromosome pairs. Twenty-two of these are autosomal chromosome pairs, while the remaining pair is sex-determining. The haploid Human genome occupies a total of just over 3 billion DNA base pairs.
Human Genome Project	The Human Genome Project was an international scientific research project with a primary goal to determine the sequence of chemical base pairs which make up DNA and to identify and map the approximately 20,000-25,000 genes of the human genome from both a physical and functional standpoint. The project began in 1990 initially headed by James D. Watson at the U.S. National Institutes of Health. A working draft of the genome was released in 2000 and a complete one in 2003, with further analysis still being published.
Axial skeleton	The Axial skeleton consists of the 80 bones in the head and trunk of the human body. It is composed of five parts; the human skull, the ossicles of the inner ear, the hyoid bone of the throat, the rib cage, and the vertebral column. The Axial skeleton and the appendicular skeleton together form the complete skeleton.
Column	The Column is a reproductive structure that can be found in several plant families: Aristolochiaceae, Orchidaceae, and Stylidiaceae. It is derived from the fusion of both male and female parts (stamens and pistil) into a single organ. This means that the style and stigma of the pistil, with the filaments and one or more anthers, are all united.
Hydrostatic skeleton	A Hydrostatic skeleton or hydroskeleton is a structure found in many cold-blooded organisms and soft-bodied animals consisting of a fluid-filled cavity, the coelom, surrounded by muscles. The pressure of the fluid and action of the surrounding muscles are used to change an organism's shape and produce movement, such as burrowing or swimming. Hydrostatic skeleton s have a role in the locomotion of echinoderms (starfish, sea urchins), cnidarians (jellyfish), annelids (earthworms), nematodes, and other invertebrates.
Skull	The Skull is a bony structure found in the head of many animals. The Skull supports the structures of the face and protects the head against injury. The Skull can be divided into two parts: the cranium and the mandible.
Sternum	The Sternum is the ventral portion of a segment of an arthropod thorax or abdomen. In insects, the sterna are usually single, large sclerites, and external.
Osteoblast	An Osteoblast is a mononucleate cell that is responsible for bone formation. Osteoblast s produce osteoid, which is composed mainly of Type I collagen. Osteoblast s are also responsible for mineralization of the osteoid matrix.
Osteoclast	An Osteoclast is a type of bone cell that removes bone tissue by removing its mineralized matrix and breaking up the organic bone . This process is known as bone resorption. Osteoclast s and osteoblasts are instrumental in controlling the amount of bone tissue: osteoblasts form bone, Osteoclast s resorb bone.
MyoD	MyoD is a protein with a key role in regulating muscle differentiation. MyoD belongs to a family of proteins known as myogenic regulatory factors . These bHLH (basic helix loop helix) transcription factors act sequentially in myogenic differentiation.

Adenosine

Adenosine is a nucleoside composed of a molecule of adenine attached to a ribose sugar molecule (ribofuranose) moiety via a β-N_9-glycosidic bond.

Adenosine plays an important role in biochemical processes, such as energy transfer--as Adenosine triphosphate (ATP) and Adenosine diphosphate (ADP)--as well as in signal transduction as cyclic Adenosine monophosphate, cAMP. It is also an inhibitory neurotransmitter, believed to play a role in promoting sleep and suppressing arousal, with levels increasing with each hour an organism is awake.

Adenosine is an endogenous purine nucleoside that modulates many physiological processes.

Deoxyadenosine diphosphate

Deoxyadenosine diphosphate is a derivative of the common nucleic acid ATP in which the -OH (hydroxyl) group on the 2' carbon on the nucleotide's pentose has been removed (hence the deoxy- part of the name.) Additionally, the diphosphate of the name indicates that one of the phosphoryl groups of ATP has been removed, most likely by hydrolysis.

Deoxyadenosine diphosphate would be abbreviated dADP.

Myofibril

Myofibril s (obsolete term: sarcostyles) are cylindrical organelles. They are found within muscle cells. They are bundles of actomyosin filaments that run from one end of the cell to the other and are attached to the cell surface membrane at each end.

Myosin

Myosin s are a large family of motor proteins found in eukaryotic tissues. They are responsible for actin-based motility.

> 'The term Myosin was originally used to describe a group of similar, but nonidentical, ATPases found in striated and smooth muscle cells.' From Pollard and Korn, 1973

Following the discovery, by Pollard and Korn, of enzymes with Myosin like function in Acanthamoeba castellanii, a large number of divergent Myosin genes have been discovered throughout eukaryotes.

Reticulum

The Reticulum is the second chamber in the alimentary canal of a ruminant animal. Anatomically it is considered the smaller half of the reticulorumen along with the rumen.

The Reticulum is colloqially referred to as the honeycomb.

Sarcolemma

The Sarcolemma is the cell membrane of a muscle cell (skeletal, cardiac and smooth muscle.) It consists of a true cell membrane, called the plasma membrane, and an outer coat made up of a thin layer of polysaccharide material that contains numerous thin collagen fibrils. At each end of the muscle fiber, this surface layer of the Sarcolemma fuses with a tendon fiber, and the tendon fibers in turn collect into bundles to form the muscle tendons that then insert into bones.

Sarcomere

A Sarcomere is the basic unit of a muscle's cross-striated myofibril. Sarcomere s are multi-protein complexes composed of three different filament systems.

· The thick filament system is composed of myosin protein which is connected from the M-line to the Z-disc by titin. It also contains myosin-binding protein C which binds at one end to the thick filament and the other to Actin.
· The thin filaments are assembled by actin monomers bound to nebulin, which also involves tropomyosin .
· Nebulin and titin give stability and structure to the Sarcomere

A muscle cell from a biceps may contain 100,000 Sarcomere s. The myofibrils of smooth muscle cells are not arranged into Sarcomere s.

Bursa

A Bursa is a small fluid-filled sac lined by synovial membrane with an inner capillary layer of slimy fluid . It provides a cushion between bones and tendons and/or muscles around a joint. This helps to reduce any friction between the bones and allows free movement.

Joint

Joint is slang for a cigarette rolled using cannabis. Rolling papers are the most common rolling medium among industrialized countries, however brown paper, cigarettes with the tobacco removed, and newspaper are commonly used throughout the developing world. Modern papers are now commonly made from a wide variety of materials including rice, hemp, and flax.

Meniscus

In anatomy, a Meniscus is a crescent-shaped fibrocartilaginous structure present in the knee, acromioclavicular, sternoclavicular, and temporomandibular joints that, in contrast to articular disks, only partly divides a joint cavity. A small Meniscus also occurs in the radio-carpal joint.

It usually refers to either of two specific parts of cartilage of the knee: The lateral and medial menisci.

Rabbits

In Australia, rabbits are the most serious mammalian pests, an invasive species whose destruction of habitats is responsible for the extinction or major decline of many native animals such as the Western Quoll. Annually, European rabbits cause millions of dollars of damage to crops. An erosion gully in South Australia caused by rabbits.

The effect of rabbits on the ecology of Australia has been devastating since their arrival from Europe in the 18th century.

Cell — The Cell is the structural and functional unit of all known living organisms. It is the smallest unit of an organism that is classified as living, and is often called the building block of life. Some organisms, such as most bacteria, are unicellular (consist of a single Cell.)

Chlamydia — Chlamydia refers to a genus of bacteria that are obligate intracellular parasites (organisms.) Many of the Chlamydia species are pathogenic. (disease-causing.)

Horse murders — The Horse murders scandal was a form of insurance fraud in the United States in which expensive horses, many of them show jumpers, were insured against death, accident and then killed to collect the insurance money. It is not known how many horses were murdered between the mid 1970s and the mid-1990s, when a Federal Bureau of Investigation (FBI) investigation brought the horse killings to light, but the number is thought to be well over 50, and may have been as high as 100. In addition, in 1977, the heiress Helen Brach disappeared and was presumed by law enforcement agents to have been murdered by the perpetrators of these crimes, because she threatened to report their criminal activity to authorities; continuing investigations into Brach's death began to uncover the insurance fraud in the 1990s.

Ascaris — Ascaris is a genus of parasitic nematode worms known as the giant intestinal roundworms. One species, A. suum, typically infects pigs, while another, A. lumbricoides, affects human populations, typically in sub-tropical and tropical areas with poor sanitation. A. lumbricoides is the largest intestinal roundworm and is the most common helminth infection of humans worldwide, an infection known as ascariasis.

Ascaris lumbricoides — Ascaris lumbricoides is the member of the Ascaris family responsible for the disease ascariasis.

It can reach a length of up to 35 cm.

Ascaris lumbricoides, or 'roundworm', infections in humans occur when an ingested infective egg releases a larval worm that penetrates the wall of the duodenum and enters the bloodstream.

Genome — In classical genetics, the Genome of a diploid organism including eukarya refers to a full set of chromosomes or genes in a gamete; thereby, a regular somatic cell contains two full sets of Genome s. In haploid organisms, including bacteria, archaea, viruses, and mitochondria, a cell contains only a single set of the Genome usually in a single circular or contiguous linear DNA (or RNA for retroviruses.) In modern molecular biology the Genome of an organism is its hereditary information encoded in DNA (or, for retroviruses, RNA.)

Genome Project — Genome project s are scientific endeavours that ultimately aim to determine the complete genome sequence of an organism (be it an animal, a plant, a fungus, a bacterium, an archaean, a protist or a virus.) The genome sequence for any organism requires the DNA sequences for each of the chromosomes in an organism to be determined. For bacteria, which usually have just one chromosome, a Genome project will aim to map the sequence of that chromosome.

Human — A Human is a member of a species of bipedal primates in the family Hominidae . DNA and fossil evidence indicates that modern Human s originated in east Africa about 200,000 years ago. When compared to other animals and primates, Human s have a highly developed brain, capable of abstract reasoning, language, introspection and problem solving.

Human Genome	The Human genome is the genome of Homo sapiens, which is stored on 23 chromosome pairs. Twenty-two of these are autosomal chromosome pairs, while the remaining pair is sex-determining. The haploid Human genome occupies a total of just over 3 billion DNA base pairs.
Human Genome Project	The Human Genome Project was an international scientific research project with a primary goal to determine the sequence of chemical base pairs which make up DNA and to identify and map the approximately 20,000-25,000 genes of the human genome from both a physical and functional standpoint. The project began in 1990 initially headed by James D. Watson at the U.S. National Institutes of Health. A working draft of the genome was released in 2000 and a complete one in 2003, with further analysis still being published.
Ovary	In the flowering plants, an Ovary is a part of the female reproductive organ of the flower or gynoecium. Specifically, it is the part of the carpel which holds the ovule(s) and is located above or below or at the point of connection with the base of the petals and sepals. In this picture of a zucchini the petals and sepals are above the Ovary and such a flower is said to have an inferior Ovary; also referred to as epigynous.
Zygote	A zygote is a term in Developmental biology used to describe the first stage of a new unique organism when it consists of just a single cell. The term is also used more loosely to refer to the group of cells formed by the first few cell divisions, although this is properly referred to as a blastomere. A zygote is usually produced by a fertilization event between two haploid cells - an ovum from a female and a sperm cell from a male - which combine to form the single diploid cell.
Copper	Copper is a chemical element with the symbol Cu and atomic number 29. It is a ductile metal with very high thermal and electrical conductivity. Pure Copper is rather soft and malleable and a freshly-exposed surface has a pinkish or peachy color.
Mammal	Mammal s (formally mammal ia) are a class of vertebrate animals whose females are characterized by the possession of mammary glands while both males and females are characterized by sweat glands, hair, three middle ear bones used in hearing, and a neocortex region in the brain. mammal s are divided into three main categories depending how they are born. These categories are, monotremes, marsupials and placentals.
Membrane	A Membrane is a layer of material which serves as a selective barrier between two phases and remains impermeable to specific particles, molecules, or substances when exposed to the action of a driving force. Some components are allowed passage by the Membrane into a permeate stream, whereas others are retained by it and accumulate in the retentate stream. Membrane s can be of various thickness, with homogeneous or heterogeneous structure.
Platypus	The Platypus is a semi-aquatic mammal endemic to eastern Australia, including Tasmania. Together with the four species of echidna, it is one of the five extant species of monotremes, the only mammals that lay eggs instead of giving birth to live young. It is the sole living representative of its family (Ornithorhynchidae) and genus (Ornithorhynchus), though a number of related species have been found in the fossil record.

Penis

The Penis of the Arachnida is an intromitent organ, present exclusively in the order Opiliones. It consists of a long shaft, the truncus, and a terminal capsule called glans, containing the stylus and ejaculatory duct. It may have from one to three muscles, or none in the specialized lineage Grassatores, where the Penis is operated by haemolymph pressure.

Edward

The Edward mango is a monoembryonic mango cultivar grown predominantly in Florida. It is considered by many to be among the finest tasting mangoes in the world; however, its poor yields have restrained the Edward from developing into a commercially significant variety.

The Edward was first propagated in the 1920s by Edward Simmonds of the Plant Introduction Garden in Miami, Florida and is believed to be a hybrid cross of Haden and Carabao mango cultivars.

Binary fission

Binary fission is the form of asexual reproduction and cell division used by all prokaryotic and some eukaryotic organisms. This process results in the reproduction of a living prokaryotic cell by division into two parts which each have the potential to grow to the size of the original cell.

Mitosis and cytokinesis are not the same as Binary fission.

Vesicle

A Vesicle is a small bubble of liquid within a cell. More technically, a Vesicle is a small, intracellular, membrane-enclosed sac that stores or transports substances within a cell. Vesicles form naturally because of the properties of lipid membranes

Fimbria

In bacteriology, Fimbria is a proteinaceous appendage in many gram-negative and gram positive bacteria that is thinner and shorter than a flagellum. This appendage ranges from 3-10 nanometers in diameter and can be up to several micrometers long. Fimbriae are used by bacteria to adhere to one another and to adhere to animal cells, and some inanimate objects.

Follicle-stimulating hormone

Follicle-stimulating hormone is a hormone synthesized and secreted by gonadotropes in the anterior pituitary gland. FSH regulates the development, growth, pubertal maturation, and reproductive processes of the human body. FSH and Luteinizing hormone (LH) act synergistically in reproduction.

Luteinizing hormone

Luteinizing hormone is a hormone produced by the anterior pituitary gland.

· In the female, an acute rise of Luteinizing hormone - the Luteinizing hormone surge - triggers ovulation.

· In the male, where Luteinizing hormone had also been called Interstitial Cell Stimulating Hormone , it stimulates Leydig cell production of testosterone.

Luteinizing hormone is a glycoprotein. Each monomeric unit is a sugar-like protein molecule; two of these make the full, functional protein.

Follicle

In botany, a Follicle is a dry unilocular many-seeded fruit formed from one carpel and dehiscing by the ventral suture in order to release seeds, such as in larkspur, magnolia, banksia, peony and milkweed.

It is rare to meet with a solitary Follicle forming the fruit. There are usually several aggregated together, either in a whorl on a shortened receptacle, as in hellebore, aconite, larkspur, columbine or the family Crassulaceae, or in a spiral manner on an elongated receptacle, as in Magnolia and Banksia.

Escherichia	Escherichia is a genus of Gram-negative, non-spore forming, facultatively anaerobic, rod-shaped bacteria from the family Enterobacteriaceae. Inhabitants of the gastrointestinal tracts of warm-blooded animals, Escherichia species provide a portion of the microbially-derived vitamin K for their host. While many Escherichia are harmless commensals, particular strains of some species are human pathogens, and are known as the most common cause of urinary tract infections, significant sources of gastrointestinal disease, ranging from simple diarrhea to dysentery-like conditions, as well as a wide-range of other pathogenic states.
Escherichia coli	Escherichia coli , is a Gram negative bacterium that is commonly found in the lower intestine of warm-blooded organisms . Most E. coli strains are harmless, but some, such as serotype O157:H7, can cause serious food poisoning in humans, and are occasionally responsible for costly product recalls. The harmless strains are part of the normal flora of the gut, and can benefit their hosts by producing vitamin K_2, or by preventing the establishment of pathogenic bacteria within the intestine.
Birth control	Birth control is a regimen of one or more actions, devices, sexual practices, or medications followed in order to deliberately prevent or reduce the likelihood of pregnancy or childbirth. There are three main routes to preventing or ending pregnancy: the prevention of fertilization of the ovum by sperm cells ('contraception'), the prevention of implantation of the blastocyst ('contragestion'), and the chemical or surgical induction of abortion of the developing embryo or, later, fetus. In common usage, term 'contraception' is often used for both contraception and contragestion.
Blood type	A Blood type is a classification of blood based on the presence or absence of inherited antigenic substances on the surface of red blood cells These antigens may be proteins, carbohydrates, glycoproteins depending on the blood group system, and some of these antigens are also present on the surface of other types of cells of various tissues. Several of these red blood cell surface antigens, that stem from one allele, collectively form a blood group system.
Cap	Adenylate CAP is an actin-binding protein that was originally identified as a binding partner for adenylate cyclase. It binds actin monomers and sequesters them from the polymerization process. The yeast ortholog of CAP is called Srv2.
Leptin	Leptin is a 16 kDa protein hormone that plays a key role in regulating energy intake and energy expenditure, including appetite and metabolism. Leptin is one of the most important adipose derived hormones. The Ob gene (Ob for obese, Lep for Leptin) is located on chromosome 7 in humans.
Helicobacter	Helicobacter is a genus of Gram-negative bacteria possessing a characteristic helix shape. They were initially considered to be members of the Campylobacter genus, but since 1989 they have been grouped in their own genus. Some species have been found living in the lining of the upper gastrointestinal tract, as well as the liver of mammals and some birds..
Helicobacter pylori	Helicobacter pylori is a Gram-negative, microaerophilic bacterium that inhabits various areas of the stomach and duodenum. It causes a chronic low-level inflammation of the stomach lining and is strongly linked to the development of duodenal and gastric ulcers and stomach cancer. Over 80% of individuals infected with the bacterium are asymptomatic.

Gamete	A Gamete is a cell that fuses with another Gamete during fertilization (conception) in organisms that reproduce sexually. In species that produce two morphologically distinct types of Gamete s, and in which each individual produces only one type, a female is any individual that produces the larger type of Gamete -- called an ovum (or egg) -- and a male produces the smaller tadpole-like type -- called a sperm. This is an example of anisogamy or heterogamy, the condition wherein females and males produce Gamete s of different sizes (this is the case in humans; the human ovum is approximately 20 times larger than the human sperm cell.)
Crassulacean acid metabolism	Crassulacean acid metabolism is an elaborate carbon fixation pathway in some plants. These plants fix carbon dioxide during the night, storing it as the four carbon acid malate. The CO_2 is released during the day, where it is concentrated around the enzyme RuBisCO, increasing the efficiency of photosynthesis.
Candida	Candida is a genus of yeasts. Many species of this genus are endosymbionts of animal hosts including humans. While usually living as commensals, some Candida species have the potential to cause disease.
Syphilis	Syphilis is a sexually transmitted disease caused by the spirochetal bacterium Treponema pallidum subspecies pallidum. The route of transmission of Syphilis is almost always through sexual contact, although there are examples of congenital Syphilis via transmission from mother to child in utero. The signs and symptoms of Syphilis are numerous; before the advent of serological testing, precise diagnosis was very difficult.
Cleavage	In embryology, Cleavage is the division of cells in the early embryo. The zygotes of many species undergo rapid cell cycles with no significant growth, producing a cluster of cells the same size as the original zygote. The different cells derived from Cleavage are called blastomeres and form a compact mass called the morula.
Gonadotropin	Human Menopausal Gonadotropin s are protein hormones secreted by gonadotrope cells of the pituitary gland of vertebrates.
Human chorionic gonadotropin	Human chorionic gonadotropin is a glycoprotein hormone produced in pregnancy that is made by the developing embryo soon after conception and later by the syncytiotrophoblast (part of the placenta.) Its role is to prevent the disintegration of the corpus luteum of the ovary and thereby maintain progesterone production that is critical for a pregnancy in humans. hCG may have additional functions; for instance, it is thought that hCG affects the immune tolerance of the pregnancy.
Morula	A Morula is an embryo at an early stage of embryonic development, consisting of cells (called blastomeres) in a solid ball contained within the zona pellucida. The Morula is produced by embryonic cleavage, the rapid division of the zygote. After reaching the 16-cell stage, the cells of the Morula differentiate.
Nereis	Nereis is a genus of polychaete worms in the family Nereidae. It comprises many species, most of which are marine, including the sandworm (Nereis virens) and the common clam worm (Nereis succinea.) Nereis possess setae and parapodia for locomotion.

Endoderm

Endoderm, (sometimes called Entoderm) is one of the germ layers formed during animal embryogenesis. Cells migrating inward along the archenteron from the inner layer of the gastrula, which develops into the Endoderm.

The Endoderm consists at first of flattened cells, which subsequently become columnar.

Gastrulation

Gastrulation is a phase early in the development of animal embryos, during which the morphology of the embryo is dramatically restructured by cell migration. Gastrulation varies in different phyla. Gastrulation is followed by organogenesis, when individual organs develop within the newly formed germ layers.

Germ

The germ of a cereal is the reproductive part that germinates to grow into a plant; it is the embryo of the seed. Along with bran, germ is often a by-product of the milling that produces refined grain products. Cereal grains and their components, such as wheat germ, rice bran, and maize may be used as a source from which vegetable oil is extracted, or used directly as a food ingredient.

Germ layer

A Germ layer is a group of cells, formed during animal embryogenesis. Germ layer s are particularly pronounced in the vertebrates; however, all animals more complex than sponges produce two or three primary tissue layers (sometimes called primary Germ layer s.) Animals with radial symmetry, like cnidarians, produce two Germ layer s (the ectoderm and endoderm) making them diploblastic.

Mesoderm

One of the three germ layers found in the embryos of animals more complex than cnidarians, making them triploblastic. Mesoderm forms in the embryo during gastrulation when some of the cells migrating inward to form the endoderm, produce an additional layer that lies between the endoderm and the ectoderm.

Mesoderm is found in all large, complex animals, and allows the formation of a coelom, which allows more room for independent growth of the body organs.

Neural tube

In the developing vertebrate, the Neural tube is the embryo's precursor to the central nervous system, which comprises the brain and spinal cord. The neural groove gradually deepens as the neural folds become elevated, and ultimately the folds meet and coalesce in the middle line and convert the groove into a closed tube, the Neural tube or neural canal (which strictly speaking is the center of the Neural tube), the ectodermal wall of which forms the rudiment of the nervous system.

There are 2 ways in which the Neural tube develops: Primary neurulation and Secondary neurulation.

No-till farming

No-till farming is a way of growing crops from year to year without disturbing the soil through tillage. No-till increases the amount of water in the soil, decreases erosion, increases the amount and variety of life in and on the soil and it increases herbicide usage.

Producing crops usually involves regular tilling that agitates the soil in various ways, usually with tractor-drawn implements.

Bud

In botany, a Bud is an undeveloped or embryonic shoot and normally occurs in the axil of a leaf or at the tip of the stem. Once formed, a Bud may remain for some time in a dormant condition, or it may form a shoot immediately.

The Bud s of many woody plants, especially in temperate or cold climates, are protected by a covering of modified leaves called scales which tightly enclose the more delicate parts of the Bud

Umbilical cord	In placental mammals, the Umbilical cord is the connecting cord from the developing embryo or fetus to the placenta. During prenatal development, the Umbilical cord comes from the same zygote as the fetus and normally contains two arteries and one vein, buried within Wharton's jelly. The umbilical vein supplies the fetus with oxygenated, nutrient-rich blood from the placenta.
Placenta	The Placenta is an organ unique to mammals that connects the developing fetus to the uterine wall. The Placenta supplies the fetus with oxygen and food, and allows fetal waste to be disposed of via the maternal kidneys. The word Placenta comes from the Latin for cake, from Greek plakóenta/plakoúnta, accusative of plakóeis/plakoús - πλακÏŒεις, πλακοÏ ς, 'flat, slab-like', referring to its round, flat appearance in humans.
Thalidomide	Thalidomide is a sedative-hypnotic, and multiple myeloma medication. The drug is a potent teratogen in rabbits and primates including humans: severe birth defects may result if the drug is taken during pregnancy. Thalidomide was sold in a number of countries across the world from 1957 until 1961 when it was withdrawn from the market after being found to be the cause of what has been called 'the biggest medical tragedy of modern times'.
Oxytocin	Oxytocin is a mammalian hormone that also acts as a neurotransmitter in the brain. It is best known for its roles in female reproduction: it is released in large amounts after distension of the cervix and vagina during labor, and after stimulation of the nipples, facilitating birth and breastfeeding, respectively. Recent studies have begun to investigate Oxytocin's role in various behaviors, including orgasm, social recognition, pair bonding, anxiety, trust, love, and maternal behaviors.

Population	In biology, a population is the collection of inter-breeding organisms of a particular species; in sociology, a collection of human beings. Individuals within a population share a factor may be reduced by statistical means, but such a generalization may be too vague to imply anything. Demography is used extensively in marketing, which relates to economic units, such as retailers, to potential customers.
Clostridium	Clostridium is a genus of Gram-positive bacteria, belonging to the Firmicutes. They are obligate anaerobes capable of producing endospores. Individual cells are rod-shaped, which gives them their name, from the Greek.kloster or spindle.
Punnett square	The Punnett square is a diagram that is used to predict the outcome of a particular cross or breeding experiment. It is named after Reginald C. Punnett, who devised the approach, and is used by biologists to determine the probability of an offspring having a particular genotype. The Punnett square is a summary of every possible combination of one maternal allele with one paternal allele for each gene being studied in the cross.
Decomposers	Decomposers are organisms that consume dead or decaying organisms, and, in doing so, carry out the natural process of decomposition. Like herbivores and predators, Decomposers are heterotrophic, meaning that they use organic substrates to get their energy, carbon and nutrients for growth and development. Decomposers use deceased organisms and non-living organic compounds as their food source.
Rat	Rat s are various medium-sized, long-tailed rodents of the superfamily Muroidea. 'True Rat s' are members of the genus Rat tus, the most important of which to humans are the black Rat Rat tus Rat tus, and the brown Rat Rat tus norvegicus. Many members of other rodent genera and families are also called Rat s and share many characteristics with true Rat s.
Predation	In ecology, Predation describes a biological interaction where a predator (an organism that is hunting) feeds on its prey, the organism that is attacked. Predators may or may not kill their prey prior to feeding on them, but the act of Predation always results in the death of the prey. The other main category of consumption is detritivory, the consumption of dead organic material (detritus.)
Leukemia inhibitory factor	Leukemia inhibitory factor an interleukin 6 class cytokine, is a chemical in cells that affects their growth and development. Leukemia inhibitory factor derives its name from its ability to induce the terminal differentiation of myeloid leukaemic cells. Other properties attributed to the cytokine include: the growth promotion and cell differentiation of different types of target cells, influence on bone metabolism, cachexia, neural development, embryogenesis and inflammation.
NADPH	Nicotinamide adenine dinucleotide phosphate ($NADP^+$, in older notation triphosphopyridine nucleotide, TPN) is used in anabolic reactions, such as lipid and nucleic acid synthesis, which require NADPH as a reducing agent. NADPH is the reduced form of $NADP^+$, and $NADP^+$ is the oxidized form of NADPH. NADP+ differs from NAD+ by the presence in NADP+ of an additional phosphate group on the 2' position of the ribose ring that carries the adenine moiety. In chloroplasts, NADP is reduced by ferredoxin-NADP+ reductase in the last step of the electron chain of the light reactions of photosynthesis.
Family	In biological classification, Family is

· a taxonomic rank. Other well-known ranks are life, domain, kingdom, phylum, class, order, genus, and species, with Family fitting between order and genus. As for the other well-known ranks, there is the option of an immediately lower rank, indicated by the prefix sub-: subfamily .
· a taxonomic unit, a taxon, in that rank. In that case the plural is families

Example: 'Walnuts and Hickories belong to the Walnut Family.

What does and does not belong to each Family is determined by a taxonomist. Similarly for the question if a particular Family should be recognized at all. Often there is no exact agreement, with different taxonomists each taking a different position.

Photosynthesis	Photosynthesis is a process that converts carbon dioxide into organic compounds, especially sugars, using the energy from sunlight. Photosynthesis occurs in plants, algae, and many species of Bacteria, but not in Archaea. Photosynthetic organisms are called photoautotrophs, since it allows them to create their own food.

Edward

The Edward mango is a monoembryonic mango cultivar grown predominantly in Florida. It is considered by many to be among the finest tasting mangoes in the world; however, its poor yields have restrained the Edward from developing into a commercially significant variety.

The Edward was first propagated in the 1920s by Edward Simmonds of the Plant Introduction Garden in Miami, Florida and is believed to be a hybrid cross of Haden and Carabao mango cultivars.

Escherichia

Escherichia is a genus of Gram-negative, non-spore forming, facultatively anaerobic, rod-shaped bacteria from the family Enterobacteriaceae. Inhabitants of the gastrointestinal tracts of warm-blooded animals, Escherichia species provide a portion of the microbially-derived vitamin K for their host.

While many Escherichia are harmless commensals, particular strains of some species are human pathogens, and are known as the most common cause of urinary tract infections, significant sources of gastrointestinal disease, ranging from simple diarrhea to dysentery-like conditions, as well as a wide-range of other pathogenic states.

Escherichia coli

Escherichia coli , is a Gram negative bacterium that is commonly found in the lower intestine of warm-blooded organisms . Most E. coli strains are harmless, but some, such as serotype O157:H7, can cause serious food poisoning in humans, and are occasionally responsible for costly product recalls. The harmless strains are part of the normal flora of the gut, and can benefit their hosts by producing vitamin K_2, or by preventing the establishment of pathogenic bacteria within the intestine.

Nereis

Nereis is a genus of polychaete worms in the family Nereidae. It comprises many species, most of which are marine, including the sandworm (Nereis virens) and the common clam worm (Nereis succinea.) Nereis possess setae and parapodia for locomotion.

Nitrogen

Nitrogen is a chemical element that has the symbol N and atomic number 7 and atomic mass 14.00674 u. Elemental Nitrogen is a colorless, odorless, tasteless and mostly inert diatomic gas at standard conditions, constituting 78% by volume of Earth's atmosphere.

Many industrially important compounds, such as ammonia, nitric acid, organic nitrates , and cyanides, contain Nitrogen.

Carbon

Carbon is the chemical element with symbol C and atomic number 6. As a member of group 14 on the periodic table, it is nonmetallic and tetravalent--making four electrons available to form covalent chemical bonds. There are three naturally occurring isotopes, with ^{12}C and ^{13}C being stable, while ^{14}C is radioactive, decaying with a half-life of about 5730 years.

Coevolution

In a broad sense, biological coevolution is 'the change of a biological object triggered by the change of a related object'. coevolution can occur at multiple levels of biology: it can be as microscopic as correlated mutations between amino acids in a protein, or as macroscopic as covarying traits between different species in an environment. Each party in a coevolution ary relationship exerts selective pressures on the other, thereby affecting each others' evolution.

Species

In biology, a Species is:

· a taxonomic rank (the basic rank of biological classification) or
· a unit at that rank

There are many definitions of what kind of unit a Species is (or should be.) A common definition is that of a group of organisms capable of interbreeding and producing fertile offspring, and separated from other such groups with which interbreeding does not (normally) happen. Other definitions may focus on similarity of DNA or morphology. Some Species are further subdivided into sub Species , and here also there is no close agreement on the criteria to be used.

Pioneer species

Pioneer species are species which colonize previously uncolonized land, usually leading to ecological succession. Since uncolonized land usually has thin, poor quality soils with few nutrients, Pioneer species are typically very hearty plants with adaptations such as long roots, root nodes containing nitrogen-fixing bacteria, and leaves that employ transpiration.

Pioneer species are often grasses such as marram grass, which grows on sand dunes.

Horse murders

The Horse murders scandal was a form of insurance fraud in the United States in which expensive horses, many of them show jumpers, were insured against death, accident and then killed to collect the insurance money. It is not known how many horses were murdered between the mid 1970s and the mid-1990s, when a Federal Bureau of Investigation (FBI) investigation brought the horse killings to light, but the number is thought to be well over 50, and may have been as high as 100. In addition, in 1977, the heiress Helen Brach disappeared and was presumed by law enforcement agents to have been murdered by the perpetrators of these crimes, because she threatened to report their criminal activity to authorities; continuing investigations into Brach's death began to uncover the insurance fraud in the 1990s.

Predation

In ecology, Predation describes a biological interaction where a predator (an organism that is hunting) feeds on its prey, the organism that is attacked. Predators may or may not kill their prey prior to feeding on them, but the act of Predation always results in the death of the prey. The other main category of consumption is detritivory, the consumption of dead organic material (detritus.)

Chara

Chara species are multicellular and superficially resemble land plants because of stem-like and leaf-like structures. The branching system is complex with branches derived from apical cells which cut off segments at the base to form nodal and internodal cells alternately. They are typically anchored to the littoral substrate by means of branching underground rhizoids.

Virus

A virus is a microscopic infectious agent that can reproduce only inside a host cell. virus es infect all types of organisms: from animals and plants, to bacteria and archaea. Since the initial discovery of tobacco mosaic virus by Martinus Beijerinck in 1898, more than 5,000 types of virus have been described in detail, although most types of virus remain undiscovered.

NADPH

Nicotinamide adenine dinucleotide phosphate ($NADP^+$, in older notation triphosphopyridine nucleotide, TPN) is used in anabolic reactions, such as lipid and nucleic acid synthesis, which require NADPH as a reducing agent.

NADPH is the reduced form of $NADP^+$, and $NADP^+$ is the oxidized form of NADPH. NADP+ differs from NAD+ by the presence in NADP+ of an additional phosphate group on the 2' position of the ribose ring that carries the adenine moiety.

In chloroplasts, NADP is reduced by ferredoxin-NADP+ reductase in the last step of the electron chain of the light reactions of photosynthesis.

Brown tree snake
The Brown tree snake is an arboreal colubrid snake native to eastern and northern coastal Australia, Papua New Guinea, and a large number of islands in northwestern Melanesia.

This snake is infamous for being an invasive species responsible for devastating the majority of the native bird population on Guam.

The Brown tree snake preys upon birds, lizards, bats and small rodents in its native range.

Photosynthesis
Photosynthesis is a process that converts carbon dioxide into organic compounds, especially sugars, using the energy from sunlight. Photosynthesis occurs in plants, algae, and many species of Bacteria, but not in Archaea. Photosynthetic organisms are called photoautotrophs, since it allows them to create their own food.

Autotroph
An autotroph is an organism that produces complex organic compounds from simple inorganic molecules using energy from light (by photosynthesis) or inorganic chemical reactions.

autotroph s are the producers in a food chain, such as plants on land or algae in water. Bacteria which derive energy from oxidizing inorganic compounds (such as hydrogen sulfide, ammonium and ferrous iron) are chemo autotroph s, and include the lithotrophs.

Decomposers
Decomposers are organisms that consume dead or decaying organisms, and, in doing so, carry out the natural process of decomposition. Like herbivores and predators, Decomposers are heterotrophic, meaning that they use organic substrates to get their energy, carbon and nutrients for growth and development. Decomposers use deceased organisms and non-living organic compounds as their food source.

Herbivore
Herbivory is a form of predation in which an organism consumes principally autotrophs such as plants, algae and photosynthesizing bacteria. By that definition, many fungi, some bacteria, many animals, some protists and a small number of parasitic plants can be considered herbivore s. However, herbivory is generally restricted to animals eating plants.

Heterochromatin
Heterochromatin is a tightly packed form of DNA. Its major characteristic is that transcription is limited. As such, it is a means to control gene expression, through regulation of the transcription initiation.

Chromatin is found in two varieties: euchromatin and Heterochromatin.

Heterotroph
A Heterotroph is an organism that uses organic substrates to get its chemical energy for its life cycle. This contrasts with autotrophs such as plants which are able to directly use sources of energy such as light to produce organic substrates from inorganic carbon dioxide. The Cyanobacteria Synechocystis sp.

Chemical
A chemical substance is a material with a specific chemical composition.

A common example of a chemical substance is pure water; it has the same properties and the same ratio of hydrogen to oxygen whether it is isolated from a river or made in a laboratory. Some typical chemical substances are diamond, gold, salt (sodium chloride) and sugar (sucrose.)

Grazing
Grazing generally describes a type of predation in which an herbivore feeds on plants (such as grasses), and also on other multicellular autotrophs (such as algae.) Grazing differs from true predation because the organism being eaten is not killed, and it differs from parasitism as the two organisms do not live together, nor is the grazer necessarily so limited in what it can eat

Many small selective herbivores follow larger grazers, who skim off the highest, tough growth of plants exposing tender shoots.

Biomass

Biomass, is a renewable energy source, biological material derived from living such as wood, waste, and alcohol fuels. biomass is commonly plant matter grown to generate electricity or produce heat. For example, forest residues (such as dead trees, branches and tree stumps), yard clippings and wood chips may be used as biofuel.

Alagille syndrome

Alagille syndrome is a genetic disorder that affects the liver, heart, and other systems of the body. Problems associated with the disorder generally become evident in infancy or early childhood. The disorder is inherited in an autosomal dominant pattern, and the estimated prevalence of Alagille syndrome is 1 in every 100,000 live births.

Chromosome

A Chromosome is an organized structure of DNA and protein that is found in cells. It is a single piece of coiled DNA containing many genes, regulatory elements and other nucleotide sequences. Chromosome s also contain DNA-bound proteins, which serve to package the DNA and control its functions.

Phosphorus

Phosphorus is the chemical element that has the symbol P and atomic number 15. A multivalent nonmetal of the nitrogen group, Phosphorus is commonly found in inorganic phosphate rocks. Elemental Phosphorus exists in two major forms - white Phosphorus and red Phosphorus.

Rat

Rat s are various medium-sized, long-tailed rodents of the superfamily Muroidea. 'True Rat s' are members of the genus Rat tus, the most important of which to humans are the black Rat Rat tus Rat tus, and the brown Rat Rat tus norvegicus. Many members of other rodent genera and families are also called Rat s and share many characteristics with true Rat s.

RNA

Ribonucleic acid (RNA) is a biologically important type of molecule that consists of a long chain of nucleotide units. Each nucleotide consists of a nitrogenous base, a ribose sugar, and a phosphate. RNA is very similar to DNA, but differs in a few important structural details: in the cell, RNA is usually single-stranded, while DNA is usually double-stranded; RNA nucleotides contain ribose while DNA contains deoxyribose (a type of ribose that lacks one oxygen atom); and RNA has the base uracil rather than thymine that is present in DNA.

RNA is transcribed from DNA by enzymes called RNA polymerases and is generally further processed by other enzymes.

Denitrification

Denitrification is a microbially facilitated process of dissimilatory nitrate reduction that may ultimately produce molecular nitrogen (N_2) through a series of intermediate gaseous nitrogen oxide products. This respiratory process reduces oxidized forms of nitrogen in response to the oxidation of an electron donor such as organic matter. The preferred nitrogen electron acceptors in order of most to least thermodynamically favourable include: nitrate (NO_3^-), nitrite (NO_2^-), nitric oxide (NO), and nitrous oxide (N_2O.)

Nitrogen cycle

The Nitrogen cycle is the biogeochemical cycle that describes the transformations of nitrogen and nitrogen-containing compounds in nature. It is a cycle which includes gaseous components.

Earth's atmosphere is approximately 78-80% nitrogen, making it the largest pool of nitrogen.

Nitrogen fixation

Nitrogen fixation is the process by which nitrogen is taken from its relatively inert molecular form (N_2) in the atmosphere and converted into nitrogen compounds (such as ammonia, nitrate and nitrogen dioxide.) This is an essential process for life because fixed nitrogen is needed to make nucleotides which are needed to make DNA and also to make amino acids which in turn are needed to produce proteins.

Nitrogen fixation is performed naturally by a number of different prokaryotes, including bacteria, actinobacteria, and certain types of anaerobic bacteria.

Root

In vascular plants, the Root is the organ of a plant that typically lies below the surface of the soil. This is not always the case, however, since a Root can also be aerial (growing above the ground) or aerating (growing up above the ground or especially above water.) Furthermore, a stem normally occurring below ground is not exceptional either

Root nodules

Root nodules occur on the roots of plants that associate with symbiotic bacteria.

Under nitrogen limiting conditions, plants from the pea family Fabaceae form a symbiotic relationship with a host-specific strain of bacteria known as rhizobia.

Within legume nodules, nitrogen gas from the atmosphere is converted into ammonia, which is then assimilated into amino acids (the building blocks of proteins), nucleotides (the building blocks of DNA and Root nodules A as well as the important energy molecule ATP), and other cellular constituents such as vitamins, flavones, and hormones.

Carbon cycle

The Carbon cycle is the biogeochemical cycle by which carbon is exchanged among the biosphere, pedosphere, geosphere, hydrosphere, and atmosphere of the Earth.

The Carbon cycle is usually thought of as four major reservoirs of carbon interconnected by pathways of exchange. These reservoirs are:

· The plants
· The terrestrial biosphere, which is usually defined to include fresh water systems and non-living organic material, such as soil carbon.
· The oceans, including dissolved inorganic carbon and living and non-living marine biota,
· The sediments including fossil fuels.

The annual movements of carbon, the carbon exchanges between reservoirs, occur because of various chemical, physical, geological, and biological processes. The ocean contains the largest active pool of carbon near the surface of the Earth, but the deep ocean part of this pool does not rapidly exchange with the atmosphere.

Fossil fuels

Fossil fuels or mineral fuels are fuels formed by natural resources such as anaerobic decomposition of buried dead organisms. The age of the organisms and their resulting fossil fuels is typically millions of years, and sometimes exceeds 650 million years. These fuels contain high percentage of carbon and hydrocarbons.

Greenhouse

A Greenhouse is a building where plants are cultivated.

A Greenhouse is a structure with a glass or plastic roof and frequently glass or plastic walls; it heats up because incoming solar radiation from the sun warms plants, soil, and other things inside the building faster than heat can escape the structure. Air warmed by the heat from hot interior surfaces is retained in the building by the roof and wall.

Greenhouse effect

The greenhouse effect is the heating of the surface of a planet or moon due to the presence of an atmosphere containing gases that absorb and emit infrared radiation. Greenhouse gases are almost transparent to solar radiation but strongly absorb and emit infrared radiation. Thus, greenhouse gases trap heat within the surface-troposphere system.

APC

APC (adenomatosis polyposis coli) is a human gene that is classified as a tumor suppressor gene. Tumor suppressor genes prevent the uncontrolled growth of cells that may result in cancerous tumors. The protein made by the APC gene plays a critical role in several cellular processes that determine whether a cell may develop into a tumor.

Salmonella

Salmonella is a genus of rod-shaped, Gram-negative, non-spore forming, predominantly motile enterobacteria with diameters around 0.7 to 1.5 Âµm, lengths from 2 to 5 Âµm, and flagella which project in all directions (i.e. peritrichous.) They are chemoorganotrophs, obtaining their energy from oxidation and reduction reactions using organic sources and are facultative anaerobes; most species produce hydrogen sulfide, which can readily be detected by growing them on media containing ferrous sulfate, such as TSI. Most isolates exist in two phases; phase I is the motile phase and phase II the non-motile phase. Cultures that are non-motile upon primary culture may be swithched to the motile phase using a Craigie tube.

Savanna

A savanna is a tropical grassland ecosystem characterized by the trees being sufficiently small or widely spaced so that the canopy does not close. The open canopy allows sufficient light to reach the ground to support an unbroken herbaceous layer consisting primarily of C4 grasses.

Some classification systems also recognize a grassland savanna from which trees are absent.

Coral

Coral s are marine organisms from the class Anthozoa and exist as small sea anemone-like polyps, typically in colonies of many identical individuals. The group includes the important reef builders that are found in tropical oceans, which secrete calcium carbonate to form a hard skeleton.

A Coral 'head', commonly perceived to be a single organism, is formed from myriads of individual but genetically identical polyps, each polyp only a few millimeters in diameter.

Human	A Human is a member of a species of bipedal primates in the family Hominidae . DNA and fossil evidence indicates that modern Human s originated in east Africa about 200,000 years ago. When compared to other animals and primates, Human s have a highly developed brain, capable of abstract reasoning, language, introspection and problem solving.
Horse murders	The Horse murders scandal was a form of insurance fraud in the United States in which expensive horses, many of them show jumpers, were insured against death, accident and then killed to collect the insurance money. It is not known how many horses were murdered between the mid 1970s and the mid-1990s, when a Federal Bureau of Investigation (FBI) investigation brought the horse killings to light, but the number is thought to be well over 50, and may have been as high as 100. In addition, in 1977, the heiress Helen Brach disappeared and was presumed by law enforcement agents to have been murdered by the perpetrators of these crimes, because she threatened to report their criminal activity to authorities; continuing investigations into Brach's death began to uncover the insurance fraud in the 1990s.
Virus	A virus is a microscopic infectious agent that can reproduce only inside a host cell. virus es infect all types of organisms: from animals and plants, to bacteria and archaea. Since the initial discovery of tobacco mosaic virus by Martinus Beijerinck in 1898, more than 5,000 types of virus have been described in detail, although most types of virus remain undiscovered.
Decomposers	Decomposers are organisms that consume dead or decaying organisms, and, in doing so, carry out the natural process of decomposition. Like herbivores and predators, Decomposers are heterotrophic, meaning that they use organic substrates to get their energy, carbon and nutrients for growth and development. Decomposers use deceased organisms and non-living organic compounds as their food source.
Deforestation	Deforestation is the logging and/or burning of trees in a forested area. There are several reasons deforestation occurs: trees or derived charcoal can be sold as a commodity and used by humans, while cleared land is used as pasture, plantations of commodities and human settlement. The removal of trees without sufficient reforestation has resulted in damage to habitat, biodiversity loss and aridity.
Desertification	Desertification is the degradation of land in arid and dry sub-humid areas, resulting primarily from man-made activities and influenced by climatic variations. It is principally caused by overgrazing, overdrafting of groundwater and diversion of water from rivers for human consumption and industrial use, all of these processes fundamentally driven by overpopulation. A major impact of desertification is biodiversity loss and loss of productive capacity, for example, by transition from land dominated by shrublands to non-native grasslands.
APC	APC (adenomatosis polyposis coli) is a human gene that is classified as a tumor suppressor gene. Tumor suppressor genes prevent the uncontrolled growth of cells that may result in cancerous tumors. The protein made by the APC gene plays a critical role in several cellular processes that determine whether a cell may develop into a tumor.
DNA	Deoxyribonucleic acid (DNA) is a nucleic acid that contains the genetic instructions used in the development and functioning of all known living organisms and some viruses. The main role of DNA molecules is the long-term storage of information. DNA is often compared to a set of blueprints or a recipe, or a code, since it contains the instructions needed to construct other components of cells, such as proteins and RNA molecules.

Drosophila

Drosophila has long been a favorite model system for geneticists and developmental biologists studying embryogenesis. The small size, short generation time, and large brood size makes it ideal for genetic studies. Transparent embryos facilitate developmental studies.

Salmonella

Salmonella is a genus of rod-shaped, Gram-negative, non-spore forming, predominantly motile enterobacteria with diameters around 0.7 to 1.5 Âµm, lengths from 2 to 5 Âµm, and flagella which project in all directions (i.e. peritrichous.) They are chemoorganotrophs, obtaining their energy from oxidation and reduction reactions using organic sources and are facultative anaerobes; most species produce hydrogen sulfide, which can readily be detected by growing them on media containing ferrous sulfate, such as TSI. Most isolates exist in two phases; phase I is the motile phase and phase II the non-motile phase. Cultures that are non-motile upon primary culture may be swithched to the motile phase using a Craigie tube.

Irrigation

Irrigation is an artificial application of water to the soil. It is usually used to assist in growing crops in dry areas and during periods of inadequate rainfall. Additionally, irrigation also has a few other uses in crop production, which include protecting plants against frost, suppressing weed growing in rice fields and helping in preventing soil consolidation.

Leukemia inhibitory factor

Leukemia inhibitory factor an interleukin 6 class cytokine, is a chemical in cells that affects their growth and development.

Leukemia inhibitory factor derives its name from its ability to induce the terminal differentiation of myeloid leukaemic cells. Other properties attributed to the cytokine include: the growth promotion and cell differentiation of different types of target cells, influence on bone metabolism, cachexia, neural development, embryogenesis and inflammation.

Plants

Plants are living organisms belonging to the kingdom Plantae. They include familiar organisms such as trees, herbs, bushes, grasses, vines, ferns, mosses, and green algae. About 350,000 species of Plants, defined as seed Plants, bryophytes, ferns and fern allies, are estimated to exist currently.

Technology

Technology is a broad concept that deals with an animal species' ethology or behavior of usage and of knowledge of tools and crafts, and how it affects the animal species' ability to control and adapt to its environment. Technology is a term with origins in the Greek 'technologia', 'τεχνολογΐ¯α' -- 'techne', 'τΐχνη' and 'logia', 'λογΐ¯α' ('saying'.) However, a strict definition is elusive; 'Technology' can refer to material objects of use to humanity, such as machines, hardware or utensils, but can also encompass broader themes, including systems, methods of organization, and techniques.

Erosion

Erosion is the removal of solids (sediment, soil, rock and other particles) in the natural environment. It usually occurs due to transport by wind, water, or ice; by down-slope creep of soil and other material under the force of gravity; or by living organisms, such as burrowing animals, in the case of bio erosion .

Soil

As defined by J.S. Joffe in 1949, soil is a natural body consisting of layers (soil horizons) of mineral constituents of variable thicknesses, which differ from the parent materials in their morphological, physical, chemical, and mineralogical characteristics. In engineering, soil is referred to as regolith, or loose rock material. soil differs from its parent rock due to interactions between the lithosphere, hydrosphere, atmosphere, and the biosphere.

Topsoil

Topsoil is the upper, outermost layer of soil, usually the top 2 inches (5.1 cm) to 8 inches (20 cm.) It has the highest concentration of organic matter and microorganisms and is where most of the Earth's biological soil activity occurs. Plants generally concentrate their roots in and obtain most of their nutrients from this layer.

Food production

The food industry is the complex, global collective of diverse businesses that together supply much of the food energy consumed by the world population. Only subsistence farmers, those who survive on what they grow, can be considered outside of the scope of the modern food industry.

The food industry includes:

· Regulation: local, regional, national and international rules and regulations for food production and sale, including food quality and food safety, and industry lobbying activities
· Education: academic, vocational, consultancy
· Research and development: food technology
· Financial services insurance, credit
· Manufacturing: agrichemicals, seed, farm machinery and supplies, agricultural construction, etc.
· Agriculture: raising of crops and livestock, seafood
· Food processing: preparation of fresh products for market, manufacture of prepared food products
· Marketing: promotion of generic products (e.g. milk board), new products, public opinion, through advertising, packaging, public relations, etc
· Wholesale and distribution: warehousing, transportation, logistics
· Retail: supermarket chains and independent food stores, direct-to-consumer, restaurant, food services

Food industry is not a formally defined term; however, it is usually used in a broadly inclusive way to cover all aspects of food production and sale. The Food Standards Agency, a government body in the UK, describes it thus:

> '...the whole food industry - from farming and food production, packaging and distribution, to retail and catering.'

The Economic Research Service of the USDA uses the term food system to describe the same thing:

> 'The U.S. food system is a complex network of farmers and the industries that link to them.

Green revolution

Green Revolution usually refers to the transformation of agriculture that began in 1945. One significant factor in this revolution was the Mexican government's request to establish an agricultural research station to develop more varieties of wheat that could be used to feed the rapidly growing population of the country.

In 1943, Mexico imported half its wheat, but by 1956, the Green Revolution had made Mexico self-sufficient; by 1964, Mexico exported half a million tons of wheat.

Kwashiorkor

Kwashiorkor is a virulent form of childhood malnutrition characterized by edema, irritability, anorexia, ulcerating dermatoses, and an enlarged liver with fatty infiltrates. The presence of edema caused by poor nutrition defines Kwashiorkor. The cause of Kwashiorkor was thought to be due to insufficient protein consumption alone, however micronutrient and antioxidant deficiencies are now believed to play important roles.

Protein	Protein s are organic compounds made of amino acids arranged in a linear chain. The amino acids in a polymer chain are joined together by the peptide bonds between the carboxyl and amino groups of adjacent amino acid residues. The sequence of amino acids in a protein is defined by the sequence of a gene, which is encoded in the genetic code.
Shigella	Shigella is a genus of Gram-negative, non-spore forming rod-shaped bacteria closely related to Escherichia coli and Salmonella. The causative agent of human shigellosis, Shigella cause disease in primates, but not in other mammals. It is only naturally found in humans and apes.
Shigella dysenteriae	Shigella dysenteriae is a species of the rod-shaped bacterial genus Shigella. Shigella can cause shigellosis (bacillary dysentery.) Shigellae are Gram-negative, non-spore-forming, facultatively anaerobic, non-motile bacteria.
Fossil fuels	Fossil fuels or mineral fuels are fuels formed by natural resources such as anaerobic decomposition of buried dead organisms. The age of the organisms and their resulting fossil fuels is typically millions of years, and sometimes exceeds 650 million years. These fuels contain high percentage of carbon and hydrocarbons.
Cell	The Cell is the structural and functional unit of all known living organisms. It is the smallest unit of an organism that is classified as living, and is often called the building block of life. Some organisms, such as most bacteria, are unicellular (consist of a single Cell.)
Chromosome	A Chromosome is an organized structure of DNA and protein that is found in cells. It is a single piece of coiled DNA containing many genes, regulatory elements and other nucleotide sequences. Chromosome s also contain DNA-bound proteins, which serve to package the DNA and control its functions.
Hydrogen	Hydrogen is the chemical element with atomic number 1. It is represented by the symbol H. At standard temperature and pressure, Hydrogen is a colorless, odorless, nonmetallic, tasteless, highly flammable diatomic gas with the molecular formula H_2. With an atomic weight of 1.007 94 u, Hydrogen is the lightest element.
Chlamydomonas	Chlamydomonas is a genus of green alga. They are unicellular flagellates. Chlamydomonas is used as a model organism for molecular biology, especially studies of flagellar motility and chloroplast dynamics, biogenesis, and genetics.
Minerals	Minerals are required by plants as part of their food, to form their structure. The firmness of straw for example, is due to the presence in it of silica, the principal constituent of sand and flints. Potassa, soda, lime, magnesia, and phosphoric acid are contained in plants, in different proportions.
Nereis	Nereis is a genus of polychaete worms in the family Nereidae. It comprises many species, most of which are marine, including the sandworm (Nereis virens) and the common clam worm (Nereis succinea.) Nereis possess setae and parapodia for locomotion.
RNA	Ribonucleic acid (RNA) is a biologically important type of molecule that consists of a long chain of nucleotide units. Each nucleotide consists of a nitrogenous base, a ribose sugar, and a phosphate. RNA is very similar to DNA, but differs in a few important structural details: in the cell, RNA is usually single-stranded, while DNA is usually double-stranded; RNA nucleotides contain ribose while DNA contains deoxyribose (a type of ribose that lacks one oxygen atom); and RNA has the base uracil rather than thymine that is present in DNA.

	RNA is transcribed from DNA by enzymes called RNA polymerases and is generally further processed by other enzymes.
Armadillo	Armadillo s are small placental mammals, known for having a leathery armor shell. The Dasypodidae are the only surviving family in the order Cingulata, part of the superorder Xenarthra along with the anteaters and sloths. The word Armadillo is Spanish for 'little armored one'.
Biochemistry	Biochemistry is the study of the chemical processes in living organisms. It deals with the structure and function of cellular components such as proteins, carbohydrates, lipids, nucleic acids and other biomolecules. Although there are a vast number of different biomolecules many are complex and large molecules (called polymers) that are composed of similar repeating subunits (called monomers.)
Biodiversity	Biodiversity is the variation of life forms within a given ecosystem, biome, or for the entire Earth. biodiversity is often used as a measure of the health of biological systems. The biodiversity found on Earth today consists of many millions of distinct biological species, which is the product of nearly 3.5 billion years of evolution.
Horseshoe crab	The Horseshoe crab or Atlantic Horseshoe crab (Limulus polyphemus) is a marine chelicerate arthropod. Despite its name, it is more closely related to spiders, ticks, and scorpions than to crabs. Horseshoe crab s are most commonly found in the Gulf of Mexico and along the northern Atlantic coast of North America.
Leprosy	Leprosy is a chronic disease caused by the bacteria Mycobacterium leprae and Mycobacterium lepromatosis. Leprosy is primarily a granulomatous disease of the peripheral nerves and mucosa of the upper respiratory tract; skin lesions are the primary external symptom. Left untreated, Leprosy can be progressive, causing permanent damage to the skin, nerves, limbs and eyes.
Root	In vascular plants, the Root is the organ of a plant that typically lies below the surface of the soil. This is not always the case, however, since a Root can also be aerial (growing above the ground) or aerating (growing up above the ground or especially above water.) Furthermore, a stem normally occurring below ground is not exceptional either
Treatise on Invertebrate Paleontology	The Treatise on Invertebrate Paleontology published by the Geological Society of America and the University of Kansas Press, is a definitive multi-authored work of some 50 volumes, written by more than 300 paleontologists, and covering every phylum, class, order, family, and genus of fossil and extant (still living) invertebrate animals. The prehistoric invertebrates are described as to their taxonomy, morphology, paleoecology, stratigraphic and paleogeographic range. However, genera with no fossil record whatsoever have just a very brief listing.
Pest control	Pest control refers to the regulation or management of a species defined as a pest, usually because it is perceived to be detrimental to a person's health, the ecology or the economy. pest control is at least as old as agriculture, as there has always been a need to keep crops free from pests. In order to maximize food production, it is advantageous to protect crops from competing species of plants, as well as from herbivores competing with humans.

Pollinator

A Pollinator is the biotic agent (vector) that moves pollen from the male anthers of a flower to the female stigma of a flower to accomplish fertilization or syngamy of the female gamete in the ovule of the flower by the male gamete from the pollen grain. Though the terms are sometimes confused, a Pollinator is different from a pollenizer, which is a plant that is a source of pollen for the pollination process.

Plants fall into pollination syndromes that reflect the type of Pollinator being attracted.

Genome

In classical genetics, the Genome of a diploid organism including eukarya refers to a full set of chromosomes or genes in a gamete; thereby, a regular somatic cell contains two full sets of Genome s. In haploid organisms, including bacteria, archaea, viruses, and mitochondria, a cell contains only a single set of the Genome usually in a single circular or contiguous linear DNA (or RNA for retroviruses.) In modern molecular biology the Genome of an organism is its hereditary information encoded in DNA (or, for retroviruses, RNA.)

Genome Project

Genome project s are scientific endeavours that ultimately aim to determine the complete genome sequence of an organism (be it an animal, a plant, a fungus, a bacterium, an archaean, a protist or a virus.) The genome sequence for any organism requires the DNA sequences for each of the chromosomes in an organism to be determined. For bacteria, which usually have just one chromosome, a Genome project will aim to map the sequence of that chromosome.

Human Genome

The Human genome is the genome of Homo sapiens, which is stored on 23 chromosome pairs. Twenty-two of these are autosomal chromosome pairs, while the remaining pair is sex-determining. The haploid Human genome occupies a total of just over 3 billion DNA base pairs.

Human Genome Project

The Human Genome Project was an international scientific research project with a primary goal to determine the sequence of chemical base pairs which make up DNA and to identify and map the approximately 20,000-25,000 genes of the human genome from both a physical and functional standpoint.

The project began in 1990 initially headed by James D. Watson at the U.S. National Institutes of Health. A working draft of the genome was released in 2000 and a complete one in 2003, with further analysis still being published.

CPSIA information can be obtained at www.ICGtesting.com
Printed in the USA
LVOW09s1420310713

345611LV00001B/7/P